ADVANCES IN ENERGY MATERIALS

Materials Fabrication, Properties, and Applications

ADVANCES IN ENERGY MATERIALS

Materials Fabrication, Properties, and Applications

Lishin T, PhD
Maciej Jaroszewski, PhD
Sabu Thomas, PhD
Editors

AAP APPLE ACADEMIC PRESS

First edition published 2026

Apple Academic Press Inc.
1265 Goldenrod Circle, NE,
Palm Bay, FL 32905 USA

760 Laurentian Drive, Unit 19,
Burlington, ON L7N 0A4, Canada

CRC Press
2385 NW Executive Center Drive,
Suite 320, Boca Raton FL 33431

4 Park Square, Milton Park,
Abingdon, Oxon, OX14 4RN, UK

Apple Academic Press exclusively co-publishes with CRC Press, an imprint of Taylor & Francis Group, LLC

Library and Archives Canada Cataloguing in Publication

CIP data on file with Canada Library and Archives

Library of Congress Cataloging-in-Publication Data

CIP data on file with US Library of Congress

ISBN: 978-1-779643-66-7 (hbk)
ISBN: 978-1-779643-67-4 (ebk)
DOI: 10.1201/9781779643674

About the Editors

Lishin T is a research scholar at the Ultrafast Science of Quantum Materials group at IMDEA Nanoscience Foundation, Madrid, Spain. He completed his postgraduation in Physics at Pondicherry University, India. His area of research is titled "Ultrafast Nanoscale Dynamics and Control of Topological Defects in Quantum Materials." His research career began at the School of Energy Materials, Mahatma Gandhi University, India, where he worked on perovskite-based optoelectronic devices. He subsequently conducted research in physics at the Catholic University of the Sacred Heart, Italy, focusing on the development of Asynchronous Optical Sampling (ASOPS), a pump-probe spectroscopy technique. Lishin has authored several publications in the fields of materials sciences and spectroscopy. He has qualified Graduate Aptitude Test in Engineering (GATE 2019) in Physics and CSIR-UGC NET for lectureship 2019 in Physics in India. His research interest includes instrumentation, machine learning, simulation, advanced characterization techniques, energy detection and spectrum analysis.

Maciej Jaroszewski, PhD, is a Titular Professor and Head of the High Voltage Group at the Wroclaw University of Science Technology in Wroclaw, Poland. He received his MS and PhD degrees in high-voltage engineering from the same university in 1993 and 1999 respectively. In 2024, the President of the Republic of Poland awarded him the title of professor of engineering and technical sciences. Prof. Jaroszewski's roles included head, contractor/prime contractor of several national grants, and a contractor of a key project co-financed by the foundations of the European Regional Development Foundation within the framework of the Operational Programme Innovative Economy. He received the Silver Medal for Long-Term Service and several awards of the Rector of the Technical University of Science and Technology Wroclaw. His research interests include high-voltage insulation systems and techniques, HV equipment diagnostics, HV test techniques, degradation of ZnO Varistors, dielectric liquids, dielectric nanoliquids and dielectric spectroscopy. He has published several books, including *Advances in Diverse Applications of Polymer Composites: Synthesis, Application, and Characterization; Advanced Materials for Electromagnetic Shielding; Modern Physical Chemistry: Engineering Models, Materials, and Methods with Applications;* and *Engineering Technologies for Renewable and Recyclable Materials: Physical-Chemical Properties and Functional Aspects;* among others.

Sabu Thomas, PhD, DSc, FRSC
Director, School of Nanoscience and Nanotechnology, Mahatma Gandhi University, Kottayam, India;
School of Energy Materials, Mahatma Gandhi University, Kottayam, India;
Department of Chemical Sciences, University of Johannesburg, Doornfontein, Johannesburg, South Africa;
International and Inter-University Centre for Nanoscience and Nanotechnology (IIUCNN), Mahatma Gandhi University, Kottayam, India

Sabu Thomas, PhD, has PhD, DSc, and FRSC degrees from the School of Nanoscience and Nanotechnology, Mahatma Gandhi University, Kottayam, Kerala, India. He is also Senior Visiting Professor at the Department of Physics and Electronics, CHRIST (Deemed to be University), Bengaluru, Karnataka, India. He is the Chairman of the TrEST Research Park, Trivandrum, Kerala, India.

He has been awarded various national and international awards, which include a Fellowship of the Royal Society of Chemistry, London; FRSC Distinguished Professorship from Josef Stefan Institute, Slovenia; MRSI Medal Nano Tech Meda; CRSI Medal; Distinguished Faculty Award; Dr. APJ Abdul Kalam Award for Scientific Excellence (2016); Mahatma Gandhi University—Award for Outstanding Contribution in November 2016; Lifetime Achievement Award of the Malaysian Polymer Group; Indian Nano Biologists Award (2017), and Sukumar Maithy Award for the best polymer researcher in the country. He is on the list of most productive researchers in India and holds a position of No. 5. Because of his outstanding contributions to the field of nanoscience and polymer science and engineering, Prof. Thomas has been conferred a Honoris Causa (DSc) Doctorate by the University of South Brittany, Lorient, France, and the University of Lorraine, Nancy, France.

Very recently, Prof. Thomas has been awarded a Senior Fulbright Fellowship to visit 20 universities in the United States. In 2019 Professor Thomas was selected as a member of the prestigious European Academy of Sciences. In the year 2021, he was awarded the Kairali Lifetime Research Award by Kerala Government. In addition, he was cited among top 2% Scientists in India by Stanford University Professors in 2020 and 2021; he was also ranked 1st in the Kerala state, 19th in the country, 236th in Asia, and 2611 in the world by German-based European Science Evaluation Centre.

Professor Thomas has published over 1200 peer-reviewed research papers, reviews, and book chapters. He has co-edited 127 books published by Royal Society, Wiley, Woodhead, Elsevier, CRC Press, Springer, Nova, Apple Academic Press, etc. He is the inventor of 15 patents. The h index of Prof. Thomas is 156, and his work has been cited more than 100,000 times. Prof. Thomas has delivered over 400 plenary/inaugural and invited lectures in national and international meetings in over 30 countries.

His areas of expertise and experience include nanomaterials, polymer blends, fiber-filled polymer composites, and polymer nanocomposites, aging and degradation, pervaporation phenomena, sorption and diffusion, interpenetrating polymer systems, recyclability and reuse of waste plastics and rubbers, dual porous nanocomposite scaffolds for tissue engineering, polymer nanocomposites for electronic applications, water purification, and energy storage.

Contents

Contributors

K. Berlin Asha
Department of Chemistry, Lady Doak College, Madurai, Tamil Nadu, India

Bibek Bandopadhyay
International Institute of Energy Conservation, New Delhi, India

Trinath Biswal
Department of Chemistry, Veer Surendra Sai University of Technology, Burla, Odisha, India

Subhra Das
Ansyst Consulting, Gurugram, Haryana, India

Tanmay De
Department of Computer Science and Engineering, National Institute of Technology, Durgapur, West Bengal, India

Aastha S. Dutta
Department of Plastic and Polymer Engineering, Maharashtra Institute of Technology, Chhatrapati Sambhaji Nagar, Maharashtra, India

Kamal Golui
Department of Mechanical Engineering, Gargi Memorial Institute of Technology, Balarampur, Baruipur, Kolkata, West Bengal, India

Sunil Kumar
Department of Chemistry, L.N.T. College Muzaffarpur, B.R.A. Bihar University, Muzaffarpur, Bihar, India

Sowmik Maji
Department of Chemical Engineering, University College of Science, Technology & Agriculture, University of Calcutta, Kolkata, India

Tanujit Mete
Department of Mechanical Engineering, Kingston Polytechnic College, Kolkata, West Bengal, India

Amit Kumar Mishra
Department of Chemistry, Indian Institute of Technology, Kanpur, India

Deepa Naik
Department of Computer Science and Engineering, National Institute of Technology, Durgapur, West Bengal, India

Joydip Paul
Department of Mechanical Engineering, Gargi Memorial Institute of Technology, Balarampur, Baruipur, Kolkata, West Bengal, India

Rituparna Roy
Department of Chemical Engineering, University College of Science, Technology & Agriculture, University of Calcutta, Kolkata, India

Subhasis Roy
Department of Chemical Engineering, University College of Science, Technology & Agriculture, University of Calcutta, Kolkata, India

Hiranmoy Samanta
Department of Mechanical Engineering, Gargi Memorial Institute of Technology, Balarampur, Baruipur, Kolkata, West Bengal, India

S. K. Suja
Department of Chemistry, Lady Doak College, Madurai, Tamil Nadu, India

Deepu Thomas
Department of Physics, Nirmalagiri College, Kerala, India

Divya Thomas
Department of Chemistry, St. Xavier's College Thumba, Kerala, India

Emmanuel Tom
Department of Physics, Nirmalagiri College, Kerala, India

Ha Thanh Tung
Vinh Long University of Technology Education, Vietnam

Preface

It is with great pleasure that we present this compilation of chapters that made their mark at the International Conference of Education and Science (ICES) 2021. The conference brought together researchers, developers, and practitioners from around the world to discuss the latest advancements in the field of energy applications. The book, titled *Advances in Energy Materials: Materials Fabrication, Properties and Applications,* is a testament to the hard work and dedication of the contributors who have made this project possible.

The book contains nine chapters that cover a wide range of topics related to energy applications, such as perovskite solar cells, solar–thermal storage materials, metal organic frameworks for energy storage, and more. Each chapter provides a detailed analysis of the topic, highlighting the latest advancements in the field and their potential impact on the energy landscape. Together, these chapters offer a comprehensive overview of the field of energy materials and its significance in the pursuit of sustainable development.

We hope that this book will serve as a valuable resource for researchers, developers, and practitioners who are working toward creating a more sustainable and eco-friendly energy landscape. We would like to express our sincere gratitude to all the contributors who have made this project possible, and we hope that this book will inspire further research and development in the field of energy applications.

—Editors

Introduction

Energy is the driving force behind our modern world, permeating every aspect of our lives. It powers our homes, fuels our businesses, and propels our transportation systems. However, the path we have taken to harness this vital resource is far from sustainable. Our current energy system relies heavily on finite fossil fuels, exacerbating the pressing issue of climate change. To address this challenge, researchers, developers, and practitioners from around the world are hard at work, striving to create new materials that are not only more efficient but also more durable and cost-effective.

The book *Advances in Energy Materials: Materials Fabrication, Properties and Applications* is a compilation of chapters that made their mark at the International Conference of Education and Science (ICES) 2021. This global gathering, held from December 10 to 12, 2021, in the picturesque locale of Kottayam, Kerala, India, brought together some of the brightest minds in the field. The book offers an expansive exploration of energy materials, delving into theoretical foundations, modeling, and simulations of energy materials, the art of synthesizing and characterizing these materials, and the multitude of applications energy materials offer in our rapidly evolving world.

The book contains nine chapters that cover a wide range of topics related to energy applications, such as perovskite solar cells, solar–thermal storage materials, metal organic frameworks for energy storage, and more. Each chapter provides a detailed analysis of the topic, highlighting the latest advancements in the field and their potential impact on the energy landscape. Together, these chapters offer a comprehensive overview of the field of energy materials and its significance in the pursuit of sustainable development.

CHAPTER 1

Recent Advancements and Future Insight on Perovskite Solar Cells: A Review

RITUPARNA ROY, SOWMIK MAJI, and SUBHASIS ROY*

Department of Chemical Engineering, University College of Science, Technology & Agriculture, University of Calcutta, Kolkata, India

*Corresponding author

ABSTRACT

In the era of depleting fossil fuels, the world is increasingly turning to various types of renewable energy sources and and embracing technological advancements to enhance their efficiency. One of the prominent technologies is perovskite solar cells (PSCs). Since 2009, this technology has come a long way by achieving >22% energy conversion efficiency from 3.8% in its starting days. This chapter focuses on the various technological advancements made in PSC recently. This includes choosing various types of materials for the perovskite layer. This includes selecting multiple types of materials for the perovskite layer from the dimensional engineering perspective, improving metal halide perovskite as the blue light emitter or as blue perovskite light-emitting diodes, using low-temperature fabrication methods, composing various types of electrons transporting layers and hole transporting layers, and using mixed halides for greater energy conversion efficiency achievement. This chapter elaborates on the temperature sensing property through the thermally coupled and uncoupled energy levels. The newly developed eco-friendly double perovskite also attracts interest. Various materials' environmental impact

Advances in Energy Materials: Materials Fabrication, Properties and Applications.
Lishin T, Maciej Jaroszewski, and Sabu Thomas (Eds.)

DOI: 10.1201/9781779643674-1

has also been assessed in this chapter. Many papers and articles have been reviewed, including theoretical and experimental works, to give a better insight into the advancement in this one of the third-generation photovoltaic technologies.

1.1 INTRODUCTION

For the more significant part of the history of mankind, mostly fossil fuels have been the primary energy generator in day-to-day life. But with the diminishing storage of fossil fuels, researchers have been extensively studying other available energy producer sources that may be abundant in nature to meet the mundane life requirement for energy. Moreover, fossil fuels are one of the main factors behind global warming, and various kinds of pollution occur in our atmosphere. Therefore, many researchers are vying for renewable energy sources. Among multiple sources, solar irradiation energy is one of the most promising possible energy sources. Besides generating no or a lesser amount of pollution, the supply of solar energy is abundant. We are being supplied 10,000 times daily of the amount that is required. Moreover, this energy source is free of cost [1].

In the first- and second-generation solar cells, many problems have been encountered while employing them daily. In Si-based solar cells, a pure and thick layer of Si film is required, but the purification process of Si is quite costly. In second-generation solar cells, the toxicity of many metals like Cd has been the main impediment [2]. In third-generation solar cells, *dye sensitized solar cells* (DSSCs) are being developed, which employ liquid polymer organic materials with higher absorption coefficients, but this poses many problems like liquid leakage, instability, and many others [3–5]. In 1998, the first 2,2′,7,7′-tetrakis-(*N*,*N*-di-4-methoxyphenylamino)-9,9-spirobifluorene (spiro-OMeTAD) was introduced in place of liquid dye for a solid-state DSSC [6]. This may reduce the leakage problem and increase incident photon-to-electron conversion efficiency, but the cell's efficiency (1%) remains challenging. Various kinds of doping were introduced to increase efficiencies, like the Li-TFSI, but the efficiency change remained very low until the first decade of the 21st century [7].

Organometal halide as a light harvester was first introduced by Miyasaka et al. in 2009 [8]. This has opened a new gate in the field of solar cells. In 2012, perovskite was used in solar cells with spiro-OMeTAD as hole transport material (HTM) generated much greater efficiency (9.7%) [9]. Later, TiO_2 is also replaced by Al_2O_3 as a scaffold layer to increase the efficiency [10] further. Since then, various HTM and electron transport material (ETM) have been deployed to enhance the perovskite solar cell (PSC) devices' efficiency and stability.

Nowadays, metal halide perovskites (MHPs) flourish hugely in the world of light emitters. It provides us with a next-level light-emitting diode (LED). It takes away a huge research field on the academic side as well as the industrial side. It has been vigorously studied for various application fields like optoelectronics, solar cells [11, 12], photodetectors [13], photocatalyst [14], lasers [15], and LEDs [16, 17]. Various research groups reported that MHP-based LEDs have various reasons for claiming themselves as next-generation light emitters. Those reasons are that it costs low for materials, has good solution processability, has high mobility for carrying purposes, has high purity in showing color, and mainly has easy tunability on the bandgap. Various advancements in MHPs are progressing to achieve high performance in different research fields, reflected in the commercial field of LEDs, especially based on organic LEDs. Such factor as the external quantum efficiencies (EQEs) has some limitations on red and green MHP LEDs (perovskite LED (PeLEDs)), which are overcome so far that they have reached nearly their theoretical value.

In contrast, the EQE value for blue MHP LEDs is lower than the green and red PeLEDs. It is reported that the values for red LEDs are 21.3% at 649 nm, for green LEDs 23.4% at 530 nm, and for blue PeLEDs 12.1% at 488 nm and 1.12% at 445 nm, which shows approximately half the value in comparison with those red and green PeLEDs [18]. As the blue light emitter has some intrinsic limitations, like it has huge proton energy [19, 20], low stability in operation [21, 22], and unstable character of the material [23, 24], and for that reason, development of this area faces different challenges.

In the case of dielectric properties of methylammonium lead iodide ($CH_3NH_3PbI_3$) (MAPI), crystals are widely studied to acknowledge the dependency on temperature and frequency. For MAPI there is an argument on the ferroelectricity existence in temperature-dependent structural studies.

Recently, double perovskites (DP) have also enormously contributed to the section of perovskite advancement. It has some interesting properties from the chemical and physical perspectives. The most attractive property is its structural composition flexibility. It has a structure of the $A_2BB'O_6$ (where this will be further discussed). As scintillators, DP takes away most of the interest in the application part, where La_2MgTiO_6 (LMT) represents itself as the host to introduce lanthanides in the DP [25–27]. Investigation results show that LMT: Er3+ can sense temperature when synthesized by the coprecipitation method by introducing thermally coupled and uncoupled energy levels. Where scintillators spread their application in the field of medical diagnosis, the field of computed tomography, nondestructive inspections, the detection department of nuclear radiation, and the security department, scintillators are also renowned as radiosensitive luminescent materials that can convert into low-energy photons or electrical charges from high-energy photons (X-rays or -rays) [28]. The conventional scintillators are $Lu_3Al_5O_{12}:Ce^{3+}$ whereas the modern scintillators are mostly the lead halide elpasolite perovskite structure $A_2^{+1}B^{+1}C^{+3}X_6$. The new advancement in scintillating material results in a low yield of X-rays, high stability, and low toxicity, whereas conventional scintillators require sintering at a very high temperature and a long time for annealing [28–32]. The environmentally friendly DP is tungstate DP, compared with other fluorides DP, silicates DP, and sulfides DP [33, 34]. The tungstate DP is also used as ceramic in the form of Ba_2MgWO_6 (BMW), which has great potential.

1.2 FUNDAMENTALS OF PEROVSKITE SOLAR CELLS

PSCs have gained significant attention in recent years due to their exceptional photovoltaic (PV) performance and potential to revolutionize the solar energy industry. The term "perovskite" refers to a specific crystal structure, and in the context of solar cells, perovskite materials generally refer to metal halide compounds with a perovskite crystal structure. The most commonly used perovskite material for solar cells is $CH_3NH_3PbI_3$. Here are some fundamental aspects of PSCs [25–29]:

1. Crystal structure: Perovskite materials have a unique crystal structure, known as the perovskite structure, characterized by a cuboctahedral arrangement of atoms. In the case of $CH_3NH_3PbI_3$, the methylammonium (MA) ($CH_3NH_3^+$) cations are located in the corner of the cube, and the lead (Pb^{2+}) and iodide (I^-) ions form the octahedral structure in the center.

2. Light absorption: Perovskite materials are excellent light absorbers, especially in the visible and near-infrared range of the solar spectrum. When the perovskite layer absorbs photons from sunlight, they generate electron–hole pairs (excitons) due to the photoelectric effect.
3. Charge transport: To generate electricity in a solar cell, the photo-excited electron–hole pairs need to be separated and collected at the respective electrodes. Efficient charge transport is essential for this process. PSCs typically include an HTM and an ETM to facilitate the movement of holes and electrons, respectively.
4. Interfaces: PSCs are multilayered devices. The perovskite layer is sandwiched between other functional layers, such as electron and hole transport layers and electrodes. The interfaces between these layers are critical for charge extraction and avoiding the recombination of charge carriers, which can reduce the solar cell's overall efficiency.
5. Stability: One of the challenges of PSCs is their relatively poor stability, especially in the presence of moisture and high temperatures. Researchers are actively working to improve the stability of perovskite materials to ensure long-term performance and commercial viability.
6. Efficiency: PSCs have shown impressive power conversion efficiencies (PCEs) relatively quickly since their discovery. The PCEs have increased rapidly over the years and have reached values comparable to traditional silicon solar cells.
7. Low-cost fabrication: PSCs have the potential for low-cost fabrication using solution-based techniques, such as spin-coating and inkjet printing, which could enable large-scale production and reduce manufacturing costs.

It is important to note that PSC technology is still an active area of research, and improvements are being made regularly. Researchers continuously explore new materials, fabrication methods, and device architectures to enhance further PSCs' efficiency, stability, and commercial viability.

Powerful PSC may be the finest choice for commercial solar technology manufacturing. For a great reason, PSC products are also separated into two categories: normal and inverted (Figure 1.1). The two main types of structures are mesoscopic and planar; mesoscopic structures often feature an n-i-p configuration, such as compact electron transporting layer (ETL)/mesoporous and ETL/perovskite/hole transporting layer (HTL)/electrode. The perovskite material is typically placed on opaque substrates in the n-i-p configuration, including an optional mesoporous TiO_2 scaffold layer.

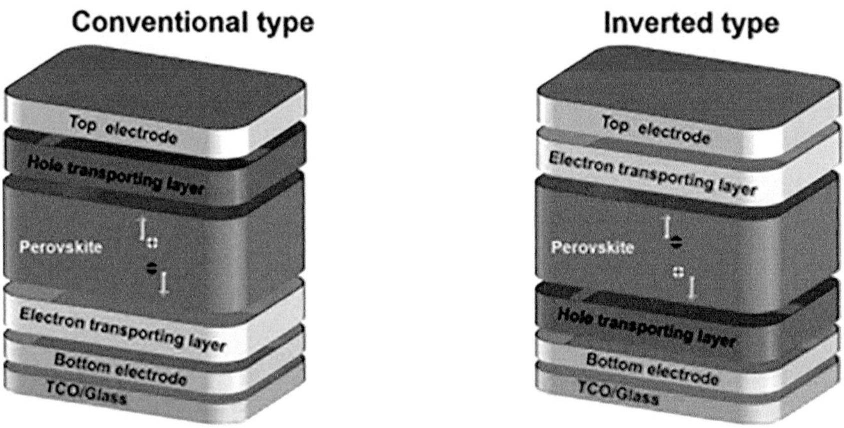

FIGURE 1.1 Two general device structures of perovskite solar cells.

Source: Reproduced with permission from Ref. [35]. https://creativecommons.org/licenses/by/4.0/

1.2.1 DEFINITION OF PEROVSKITE SOLAR CELLS AND DOUBLE PEROVSKITE SOLAR CELLS

Perovskite presents a general ABX_3 structure where A is a cation that coordinates 12 X anions, B is a cation that coordinates 6 X anions, and X is anions that can be anyone among oxygen, carbon, nitrogen, or halide. In most cases, X is usually a halide molecule due to their better electrical and magnetic properties. The most common example of perovskite is $MAPbI_3$, where MA is MA cation [36].

1.2.1.1 DOUBLE PEROVSKITE

DP material has snatched the focus of every researcher interested in the perovskite field, its structural flexibility in adopting different compositions, and its interesting physical and chemical properties.

DP is an extended part of the research field on the perovskite structure. Where perovskite has the formula of $ABO_{3,}$ DP can be represented by doubling the perovskite formula resulting in $A_2B_2O_6$ (where an organic cation is represented by A, a divalent metal ion is represented by B, and X can represent a halide). In addition, DP can also hold two different transition metals (TMs) at the B position of the structural formula of DP. If a new B' cation replaces one B cation, then the DP formula results in $A_2BB'O_6$. This type of DP shows the possibility of accommodating two different rare-earth and alkaline earth metals in the A-site, having divalent cations like Sr, Ca, and Ba on the A position and the B and B' positions. There are 3d and 4d or 5d TM ions. It shows better properties in the magnetic field than the single perovskites due to the presence of localized and delocalized 3d and 4d, or 5d TMs, respectively.

1.3 EXPERIMENTAL

1.3.1 FABRICATION METHOD OF PEROVSKITE SOLAR CELLS

There are many techniques to fabricate a PSC, but here is a simple process to fabricate a PSC using specific materials.

In PSCs, devices are made of an anode, the ETM, an active layer, the HTM, and a cathode. At first, cathode fluorine-doped tin oxide (FTO) conductive glass is taken, which has an energy level of 4.4 eV. Therefore, as ETM, a material has been chosen that has a comparable energy level with FTO. Hence, TiO_2 with 4 eV is taken as ETM. For the active layer, $MAPbI_3$ is taken, having bandgaps between (3.93 eV) and (5.43 eV). Next, for HTM, CuSCN is born with a valence band of 5.3 eV, and carbon powder and FTO are used for the anode.

First, FTO is cleaned, and with multimeter conducting side is chosen. TiO_2 is deposited using spin coating, and then at 425°C, it is annealed for 30 min. $MAPbI_3$ layer is deposited over TiO_2 film by spin coating at 1000 rpm for 30 s, and then at 120°C, it is also annealed for 30 min. Then it is cooled in an ambient atmosphere for approximately 10 min. Next, CuSCN deposition is spun coated at 500 rpm for 30 seconds, and at 80°C, it is annealed for 15 min. Then it is cooled for 5 min. On another FTO glass substrate, carbon powder is sprinkled to form a thin layer, and then both the cathode and anode are attached along with their conductive sides with deposited layers on top of each other [37].

1.4 TECHNOLOGICAL ADVANCEMENTS

HTMs and ETMs are crucial components in certain types of solar cells, particularly in PSCs and organic photovoltaics. These materials play important roles in facilitating the movement of charge carriers (electrons and holes) within the solar cell to generate electricity efficiently. In a solar cell, photons from sunlight are absorbed by the active layer (e.g., perovskite or organic materials), generating electron–hole pairs. The holes, which are the positively charged carriers, need to be efficiently transported from the active layer to the anode (positive electrode) of the solar cell. This is where HTMs come into play. HTMs have properties that allow them to accept and transport holes effectively, minimizing losses during this process. One commonly used HTM in PSCs is spiro-OMeTAD. It has been popular due to its relatively high hole mobility and stability. However, research in the field is ongoing, and new materials are being developed to improve the performance and stability of PSCs [35].

However, electrons generated in the active layer need to be efficiently transported from the active layer to the solar cell's cathode (negative electrode). This is where ETMs come into play. ETMs are materials with properties that facilitate the transport of electrons while minimizing losses during this process. In some PSCs and organic PVs, materials like fullerene derivatives (e.g., phenyl-C_{61}-butyric acid methyl ester [PCBM]) have been widely used as ETMs. However, like HTMs, research continues to develop new and improved ETMs to enhance the overall performance of solar cells [36].

For example, the device structure and energy level diagram of PSCs are shown in Figure 1.2 a and b, respectively. The perovskite-sensitized TiO_2 porous layer, compact layer, and conductive substrate are all components of the inorganic-organic hybrid heterojunction solar cells, a metal electrode, a hole-transporting layer, and a layer. The TiO_2 compact layer prevents recombination and transfers the electron. The perovskite sheet will absorb the majority of the light. The electron and hole can efficiently split due to the matched band structure. The perovskite film should be filled into a TiO_2 porous film with appropriate pore size and thickness to attain high efficiency. The perovskite-sensitized TiO_2 porous layer was coated with a thin HTL to prevent direct contact with the Au electrode.

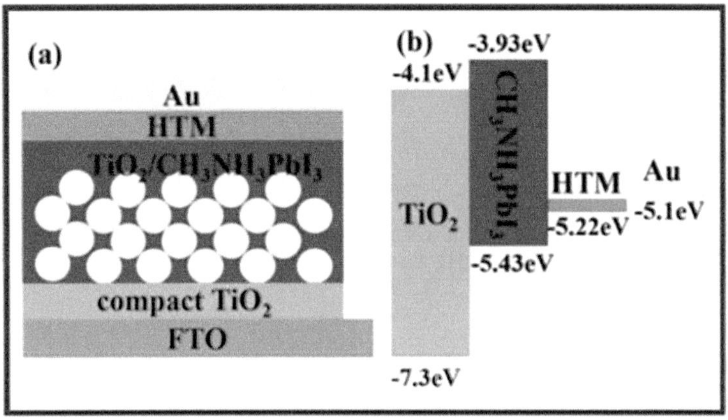

FIGURE 1.2 (a) The device structure and (b) energy level diagram of the perovskite solar cell. *Source*: Reproduced with permission from Ref. [38]. Copyright 2014. American Chemical Society

1.4.1 THROUGH COMPOSITE MATERIALS OF PSC

The most commonly used perovskite materials are $MAPbI_3$ and $MAPbBr_3$. But with single halide composition, higher PCE is far more to reach. Hence, various groups encountered mixed-halide compositions to improve PSC performance. With $MAPb(I_{1-x}Br_x)_3$ and $MAPb(I_{1-x}Cl_x)_3$ $(0 \leq x \leq 1)$, a tunable bandgap up to 3 eV can be achieved [38–40]. Moreover, mixed halide perovskite $MAPbI_{3-x}Cl_x$ has a higher diffusion length than single halide perovskite [41].

However, the main problem remains with the toxicity of Pb molecule $MAPbX_3$. Several other metals, such as Sn or Bi, have been used in place of Pb in perovskite cells [42,43]. In the case of Sn-based PSC, the stability factor comes as an impediment to commercialization, as Sn^{2+} readily oxidizes to Sn^{4+} and hence reduces the lifetime of the devices [44,45]. However, in Bi-based PSC, better stability can be achieved at the expense of the compromise of cell performance [46–48].

Recently, many researchers have used mixed cations instead of single cations in the perovskite layer. MA cation with formamidinium cation has reached PCE value up to 20.8% [49]. Cesium cation with the former two cations-based PSC devices has retained PCE of 19.5% even up to 18 weeks in ambient atmosphere. A newly developed cesium-based $CsSnGeI_3$ perovskite cell has provided better cell efficiency with PCE of 13.29% [50].

1.4.2 THROUGH HOLE-TRANSPORTING MATERIAL

In a PSC, the hole-transfer material is a very important part of the structure. The HTM contributes to PSCs in various ways: (1) increase the hole transporting efficiency owing to their high holes mobility; (2) slow down the degradation and corrosion occurring in PSCs; and (3) reduce recombination of charges by separating the contact of holes from electrons [51]. HTMs are chosen while keeping some important factors in mind: (1) possessing suitable energy levels, (2) act as an electron-blocking layer between anode and perovskite layer, (3) should have higher hole transfer efficiency, (4) must block the diffusion between perovskite layer and back electrode metals, and (5) must be less costly, possessing improved thermal stability and an impediment to degradation caused by moisture and oxygen, eco-friendly.

Previously, in highly efficient perovskite cells, perovskite layers and transporting layers were prone to degrade due to instability caused by air, moisture, light, water, and heat. The choice of HTMs is important to achieve higher efficiency and better stability. Various compounds, such as spiro-OMeTAD, PTAA [poly tri-arylamine], PEDOT: PSS [poly(3,4-ethylene dioxythiophene) polystyrene sulfonate polymer], P3HT [poly(3-hexylthiophene)], etc., are considered for HTM material. Among these, spiro-OMeTAD is chosen because of various properties it has, for example, glass transition temperature, ionization potential, absorption spectrum, solubility, and solid-state morphology [52]. Though it helped achieve more than 22% PCE for solar cells [53], it is prone to decay due to instability toward water and light. Hence, many additives were introduced to enhance the stability of pristine spiro-OMeTAD. Burschka et al. [53] introduced lithium and antimony-based salts like $Li[(CF_3SO_2)_2N]$ (Li-TFSI), but lithium aggravates the degradation of the perovskite layer. Wang et al. [54] introduced 4-*tert*-butyl pyridine to prevent the accumulation of Li-TFSI, but it also results in corrosion in the perovskite layer. Also, hydrophobic materials like tetrafluoro-tetracyanoquinodimethane [F4-TCNQ] doped spiro-OMeTAD improved the cell efficiency but failed to enhance stability [55]. Later, copper salt doped (CuSCN or CuI) spiro-OMeTAD has shown to be a better HTM material and also slows down the degradation of the perovskite layer by suppressing the aggregation of film and reducing the chances of crystallization of spiro-OMeTAD, but the efficiency achieved from this is as high as 18% [56].

Other HTM materials like PEDOT: PSS, P3HT, and PTAA have also been encountered. PEDOT film has the highest occupied molecular orbital energy level of 5.5 eV to the perovskite layer and is also a cost-efficient material but possesses instability toward the environment [57]. Hence, PSS was incorporated into it to rectify the problem. GO-modified PEDOT: PSS shows better PCE (15.34%) than PEDOT: PSS (11.90%), and also, the GO solution helps to remove the hydrophilic PSS material to enhance the moisture resistance and GO also inhibits the recombination process in the perovskite layer [58]. P3HT can also be considered an HTM material due to its higher charge transfer, low cost, and better natural stability [59]. F4-TCNQ doped P3HT showed improved PCE of 14.4%, higher than pristine P3HT (10.3%) due to enhanced conductivity and better stability in the humid atmosphere [60]. Doped with cobalt complex compounds like tris (2-(1H-pyrazol-1-yl) pyridine) cobalt(II) di[bis-(trifluoromethane) sulfonamide] (Co(II)-TFSI) in P3HT showed better conductivity, which results in higher PCE of 16.28% [61]. Slightly modified PTAA as HTM showed better open-circuit voltage and better PCE [62]. A bilayer of r-GO/ PTAA enhances the stability of the devices, and due to large perovskite grains, enhancement of light harvesting can be achieved as r-GO also can absorb near-UV lights [63].

Stability and efficiency remain the main challenges in the above-mentioned HTMs; also, they are costly to some extent. Therefore, small organic molecule-based HTMs are encountered along the route. Pyrene-based 1-(*N*,*N*-di-*p*-methoxyphenylamine) pyrene shows better charge transfer ability and achieves around 12% efficiency, comparable with spiro-OMeTAD [64]. TAPC [4,4′- cyclohexylidenebis [*N*,*N*-bis(4-methyl phenyl) benzenamine]] can be seen as a replacement for PEDOT: PSS due to its uniformity, hydrophilicity, and higher hole conductance without any doping as it achieved PCE of 18.80 % [65]. Carbazole-based 1,3,6,8-tetra (*N*,*N*-dimethoxypenylamino)-9-ethyl carbazole shows PCE of 17.8% [66]. Methoxy diphenylamine-based HTM showed a better PSC of 19.96% [67, 68].

Along with organic HTMs, various metal-based HTMs are examined thoroughly to achieve stable PSC. Among these, nickel, copper, or other TM-based HTMs are more efficient than others. Nickel oxide is a good HTM material due to its comparable HOMO energy level (5–5.6 eV) to the perovskite layer, electron leakage prevention, and optical transparency property [69]. Nickel oxide can also prevent charge recombination, thus enhancing hole transportation and being noncorrosive, making it one of the promising candidates for HTM [70]. But NiO's main challenges are lower efficiency and affinity toward contaminants, including carbonaceous materials. UV-ozone treatment can improve efficiency by reducing the efficiency [71]. Nanocrystals NiO can increase the optical path length by scattering the lights and enhancing better efficiency. Photoconductive NiO_x-Pc is inculcated by combining NiO_x with phthalocyanine, which can enhance the conductivity by more than 300%, which results in 21.68% efficiency [72]. Various additives are introduced to improve the efficiency of NiO as HTM. While 5% Cu is introduced into the NiO_x lattice, 15.4% of PCE value can be observed with V_{OC} of 1.11 V [73]. 4% Pr-doped NiO_x-based PSCs have shown better conductivity of 8.27×10^{-6} s/cm with PCE of 9.24% due to lower charge recombination. Also, 1% Zn-doped NiO_x has shown 14.72% PCE with V_{OC} of 1.02 V, which is less prone to degradation [74].

Various derivatives of Cu can also be seen as a promising HTM materials to achieve better stability of devices. CuI has a high fill factor compared to spiro-OMeTAD but lacks efficiency [75]. CuI in inverted planar heterojunction PSCs has shown better stability and performance than PEDOT:PSS due to its hydrophobicity [76]. A novel chromium–copper binary metal oxide having a wide bandgap of around 3.1 eV shows a better PCE value of 17.19% on glass [77]. Also, $CuGaO_2$ has shown better stability and a higher PCE of 18.51% [78]. $CuGaO_2$-based HTMs have shown thermal stability and resistance to moisture corrosion. These devices have retained 83% of initial efficiency in the ambient atmosphere even after 400 hours. Another copper derivative, copper phthalocyanine (CuPc) nanorods with carbon as cathode material processed at low temperature, showed an efficiency of 16.1% with a fill factor of 74% [79] due to its higher mobility of hole and longer exciton diffusion length. With F4-TCNQ-doped CuPc, a maximum of 20.16% PCE was obtained [80]. Due to its low cost, higher stability, and excellent resistance toward moisture, CuPc can be seen as a replacement for PEDOT: PSS [81].

Among the TM oxides, MoO_x can be attributed to being an impeccable candidate for HTM because of its high work function (6.9 eV), higher conductivity (1.2×10^{-7} s/cm), nontoxicity, and higher transmittance besides its stability [82–84]. Because of the electronic structure and conductivity, MoO_x has shown better hole extraction properties [85].

UV annealing of MoO_x removed contaminants, and a better PCE can be achieved (9.27% in PBDB-T: ITIC BHJ cell) [86]. The film properties of MoOx can be improved through various types of doping. Vanadium-doped MoO_x has shown lesser resistance toward hole transportation [87]. NiO_x-doped MoO_x has demonstrated a better ability to impede electrons and enhance the contact to charge transport toward the anode [88]. A composite of copper bromide and MoO_x has resulted in an enhanced fill factor of 65.20% [89].

Another metal oxide that attracted many eyes due to its smooth surface morphology and increased charge mobility is tungsten oxide (WO_x) [90]. Incorporating 10 wt% Au nanoparticles, Au-WO_x nanocomposite as an HTM has shown better surface smoothness, increasing PCE by around 60% compared to normal WO_x [91]. Silver-doped WO_x has better PCE achievement, with the highest being 4.6% [92].

1.4.3 THROUGH ELECTRON-TRANSPORTING MATERIAL

The electron-transfer material is one of the most prominent layers in PSC, which contributes to the performance enhancement and stability of the devices. The properties that made an ETM efficient are (1) the extraction of electrons from the perovskite layer and transportation of those; (2) facilitation of contact between electrode and perovskite layer; (3) prevention of hole transportation toward cathode; and (4) serve as a moisture-repellent layer to achieve better stability [53, 92, 93]. Various types of materials are experimentally applied in various devices keeping in mind some essential factors like (1) ETMs should prove sheath to the perovskite layer, thus preventing pinholes or shunting; (2) ETMs should possess higher optical transmittance and should be able to reduce optical losses; (3) ETMs must have comparable conduction band with perovskite layer; (4) possess ambient atmosphere stability; and (5) ETMs must have higher electron mobility.

Many materials have been thoroughly examined for good ETM material. Among them, TiO_2 has better performance than SnO_2, ZnO, and WoO_x. But in the case of inverted perovskite devices, which lack a dense layer of inorganic ETM to absorb sufficient solar radiation, organic ETMs have also been ventured [94, 95]. Examples of organic ETMs are fullerene or fullerene derivatives like [6,6]-PCBM and fullerene (C_{60}) [96].

Among inorganic ETMs, TiO_2 is used extensively due to its wide bandgap and preferable conduction band concerning the perovskite layer. TiO_2 possesses four phases: anatase, rutile, brookite, and TiO_2-B [7, 8]. Among these, anatase TiO_2 possesses PCE >24% [9]. Rutile $TiO_{2,}$ due to its improved charge extraction and lesser chances of recombination and acting as a reflecting layer, has shown PCE up to 20.9% [97, 98]. But because of higher photocatalytic activity and higher temperature requirements while processing, TiO_2-based devices have shown quicker degradation, resulting in instability.

SnO_2 was considered an ETM material due to its ability to block hole transportation or prevent current leakage. Also, it can be processed at low temperatures compared to TiO_2 [99–101]. It has shown higher optical transmittance and high bulk electron mobility (240 $cm^2 V^{-1} s^{-1}$) [102, 103]. Ultrafine SnO_2 nanoparticles have yielded 19.2% efficiency [104]. Various materials are doped with SnO_2 to achieve better performance. Sb-doped SnO_2 has resulted in 17.7% PCE with a better fill factor [105]. Li-doped SnO_2 has improved the PCE to 18.2% [106, 107]. Around 0.50% Nb in SnO_2 has improved the PCE to 17.57%, but Nb accumulates onto the SnO_2 surface [108]. Ta-doped SnO_2 has shown enhanced conductivity of 4.7 s/cm and reduced the decaying process. This ultimately results in improved PCE of 20.80% [109]. Many metal oxides are composited with SnO_2. TiO_2 and SnO_2 composite has shown better stability where defects passivation occurs from TiO_2/perovskite layer, and SnO_2 has also blocked hole transportation. This ETM has shown PCE of 18% [110]. In ZnO and SnO_2 composite, the injection of an electron from SnO_2 to ZnO has enhanced the $V_{OC,}$ which results in a more excellent PCE value of 19.1% [111]. It has also shown better conductivity than pristine SnO_2.

In many cases, ZnO has also been encountered as an ETM material due to its comparable conduction band and wide bandgap. It also has a higher electron mobility of 300 $cm^2V^{-1}s^{-1}$ [112]. They can be processed in low temperatures also [113]. But due to poor thermal stability and lower efficiency, they are deferred from being an ideal candidate [114, 115]. ZnO also has a high isoelectric point. Hence they attract protons from the cations, which results in the degradation of the device [116]. Although with ZnO as ETM, devices with 19.84 % PCE can be achieved [112].

Among other inorganic ETMs, $BaSnO_3$ has achieved a maximum PCE of 21.2% and, after 1000 hours of operation, has retained 93% of initial performance [117] but has a lower performance problem concerning TiO_2 and SnO_2.

In the case of inverted PSC, organic ETM molecules are introduced due to the thick layer of solar radiation compared to inorganic ETMs. Among them, fullerene-based ETMs have shown better performance compared to others. Fullerene possesses high mobility ability with better transportation. PCBM has a higher LUMO energy level in inverted PSC [118]. *N*-methyl-2-pentyl-[60] fullerene pyrrolidine has higher conductivity than PCBM and showed a better PCE of 13.83% [119]. Doping into fullerene has shown better mobility and higher performance. Fullerene doped with MA iodide has increased conductivity by 100 times [120].

1.4.4 DIFFERENT BAND ENERGY

In multilayer solar cell structures, using materials with different bandgap energies for each layer is common. The purpose of using materials with different bandgap energies is to optimize the absorption of photons across a broad range of the solar spectrum and enhance the solar cell's overall efficiency. This concept is known as a tandem or multijunction solar cell. Here is how it works [117–120]:

1. Tandem solar cells: Tandem solar cells consist of multiple photoactive layers with varying bandgap energies stacked on each other. Each layer is designed to absorb light in different parts of the solar spectrum. Typically, the higher bandgap energy layer is placed on top, and the lower bandgap layer is placed underneath.
2. Bandgap engineering: The bandgap of a material determines the range of light frequencies (or wavelengths) it can absorb. Materials with larger bandgap energies absorb higher energy (shorter wavelength) photons, while materials with smaller bandgap energies absorb lower energy (longer wavelength) photons. By carefully selecting materials with appropriate bandgap energies, it is possible to match the absorption spectra of different layers to different regions of the solar spectrum.
3. Absorption complementarity: The key advantage of using materials with different bandgap energies in tandem solar cells is that each layer can absorb a different portion of the solar spectrum. This allows more efficient utilization of incoming sunlight, as each layer converts photons from its specific spectral range into electrical current.

4. Efficient charge collection: The charge carriers (electrons and holes) generated in each layer need to be efficiently collected to produce electrical current. Typically, intermediate layers called “tunnel junctions” or “interconnects” facilitate the flow of charge carriers between the different layers in the tandem solar cell.
5. Improved efficiency: Combining materials with different bandgap energies in a tandem structure can significantly improve the solar cell's overall efficiency compared to single-junction solar cells. Tandem solar cells have the potential to achieve higher efficiencies because they can capture a larger fraction of the solar spectrum.

It is important to note that the design and optimization of tandem solar cells can be complex, and researchers continue to explore novel materials and device architectures further to enhance the efficiency and performance of multilayer solar cells. Additionally, the specific materials used and the arrangement of the layers depend on the intended application and the desired efficiency range.

1.4.5 EXCITONIC BINDING ENERGY

Excitonic binding energy is crucial in optoelectronic applications, particularly semiconductor materials used for light emission and PV devices. An exciton is an electron's bound state, and a hole is created when an electron is excited from the valence band to the conduction band. The energy required to create an exciton is known as the excitonic binding energy [92–99].

1. PV devices (solar cells): In PV devices, such as solar cells, the excitonic binding energy affects the charge separation efficiency and photocurrent generation. When photons are absorbed in the semiconductor material, electron–hole pairs (excitons) are generated. For efficient solar cell operation, these excitons need to dissociate into free charge carriers (electrons and holes) that can be collected at the respective electrodes to generate electrical current. Suppose the excitonic binding energy is too high. In that case, separating the electron–hole pairs and generating free charge carriers becomes more challenging, leading to a lower photocurrent and reduced overall efficiency of the solar cell. However, suppose the binding energy is appropriately matched with other

material properties. In that case, it can facilitate the efficient dissociation of excitons, contributing to the solar cell's higher power conversion efficiency. The excitonic binding energy in both LEDs and solar cells depends on the specific semiconductor material used. Researchers continuously explore new materials and material combinations to optimize the excitonic binding energy and other electronic properties to improve the performance of optoelectronic devices. Understanding and controlling the excitonic properties of semiconductor materials is critical for advancing optoelectronic technologies and realizing more efficient light emission and energy conversion.

2. LEDs: The excitonic binding energy is important for determining light emission efficiency in LEDs. When an electron–hole pair recombines in a direct bandgap semiconductor, it releases energy as a photon, leading to light emission. The energy of the emitted photon is ideally equal to the energy difference between the conduction band and valence band (bandgap energy). However, in practice, the emitted photon's energy is slightly lower due to the excitonic binding energy. The excitonic binding energy affects the radiative recombination rate in the LED. If the binding energy is too high, the electron–hole pairs can remain bound as excitons for more extended periods, reducing the number of radiative recombination events and decreasing the LED's efficiency. On the other hand, if the excitonic binding energy is too low, it can lead to rapid nonradiative recombination processes, resulting in reduced efficiency. Therefore, a suitable balance of excitonic binding energy is essential for optimizing the efficiency of LED devices.

1.4.6 HYBRID PEROVSKITES VERSUS METAL HALIDE PEROVSKITES

1. Hybrid perovskites have garnered significant attention for solar cells and LEDs due to their unique properties and versatility in optoelectronic applications. However, the reasons for using hybrid perovskites in solar cells and MHPs in LEDs are distinct and based on their respective characteristics. (1) Hybrid perovskites for solar cells: Hybrid perovskites, such as $CH_3NH_3PbI_3$, have become prominent materials for solar cells for several reasons [25–29]:

a. Optimal bandgap: Hybrid perovskites can be engineered to have a tunable bandgap by adjusting their chemical composition. This property allows them to effectively absorb a broad range of solar wavelengths, including visible and near-infrared light. This tunability makes them suitable for harvesting a more significant portion of the solar spectrum, leading to higher solar cell efficiency.
b. Low-cost and easy fabrication: Hybrid perovskites can be synthesized using solution-based techniques, which are relatively low-cost and compatible with large-scale manufacturing processes. These simple fabrication methods can potentially reduce the overall production costs of solar cells, making them more economically viable.
c. High carrier mobility: Hybrid perovskites exhibit high charge carrier mobility, which means that the generated charge carriers (electrons and holes) can move efficiently within the material. This property is essential for achieving high charge collection efficiency in solar cells.
d. Rapid advancements: The research and development in the field of hybrid PSCs have progressed rapidly, leading to impressive improvements in power conversion efficiency over a short period.

2. MHPs for LEDs: MHPs, such as cesium lead halides ($CsPbX_3$, where X = Cl, Br, or I), are gaining attention for their potential use in LEDs due to the following factors:

   a. Efficient light emission: MHPs can emit light efficiently when subjected to an external stimulus, making them suitable candidates for LEDs. When an electric current passes through the material, the recombination of charge carriers leads to the emission of photons.
   b. Color purity: MHPs exhibit vivid and pure colors when emitting light, making them attractive for display and lighting applications. They have shown promise in generating highly saturated and tunable colors, which is desirable for high-quality displays.
   c. Solution processing: Like hybrid perovskites, MHPs can be processed using solution-based techniques, enabling low-cost and scalable fabrication of LEDs.

d. Versatility: MHPs can be synthesized with various halide compositions, enabling tuning their emission wavelengths. This tunability allows for developing LEDs that emit light across various colors.

However, it is important to note that hybrid perovskites and MHPs are active research areas, and challenges related to stability and toxicity are being addressed to ensure their commercial viability and safe use in various applications.

1.4.7 METAL HALIDE PEROVSKITE AS BLUE LIGHT EMITTING DIODES

The newly introduced next-generation light emitter, MHP, shows attractive properties in the purity of color, high carrier mobility, flexible nature on adjusting bandgap, and cost-efficiency over the conventional LEDs. In the case of PeLEDs, they also claim themselves the same for displaying solid-state lighting. Blue PeLEDs show an efficiency value of half of the efficiency value of red and green PeLEDs. Where red PeLEDs show 21.3% at 649 nm emission wavelength and green PeLEDs at 530 nm emission wavelength show 23.4%, the blue PeLEDs at 488 nm emission wavelength results in 12.1%, and 445 nm emission wavelength results in 1.12%. Different research has been done on the advancement of PeLEDs from approximately 7 years ago to now, but nowadays, the study focuses on the development of blue PeLEDs as it faces different challenges. To address this issue on a global scale, various research teams have focused their efforts on achieving their goal, which also contributes to advancements in the display technologies of the perovskite family.. They have attempted different fabrication methods to develop the blue PeLEDs based on dimensional engineering [121, 122], additive engineering upon MHPs [123, 124], and the adjacent functional layers; there are some indirect modifications, which also include a hole and electron-transporting layers [125–127]. Here researchers only focus on two different strategies to develop blue PeLEDs, which are highly efficient, one is a mixed halide perovskite, and the other one is low-dimensional MHPs.

In the case of MHPs, one approach is tuning the electroluminescence spectrum of different metal halide LEDs, which is based on a compositional engineering basis for the X position of the MHPs of $APbX_3$, form where X represents the halide ion. Through the orbital state study of the respective elements of single halide MHPs, researchers can determine based on the electronic band structure of 3D MHPs. Therefore, partial hybridization is there, and also the spin coupling is there, 6s orbital of Pb and 5p orbital of single halide, which becomes the highest valance band (mainly preferable is p orbital of halides) and as well as the lowest conduction bands are the 6p orbital of Pb and 5p orbital of halide (mainly preferable is the p orbital of Pb) [40, 128, 129]. For the variation in X-site halide atoms like Cl, Br, and I in $APbX_3$, we can observe the difference in colors of the emitted light due to the variation in the formation of valence band maximum (VBM) and also the variation in conduction band minimum (CBM) of the $APbX_3$. MHPs containing pure chloride emit a spectrum of deep blue-violet, approximately at 409 nm emission wavelength, with a large bandgap (3.16 eV) [130, 131]. It becomes challenging to synthesize pure blue PeLEDs in the emission wavelength of 465 nm using a single halide 3D mixed halide perovskite. One can observe an effective change in structure, optical behavior, and electronic properties when halide ions are mixed with MHPs. The increasing order of ionic radius of halide ions from chlorine to iodine significantly impacts the addition of MHPs. There, one can observe the change in lattice constant having a distortion in a lattice structure, resulting in a huge change in center-to-center distance between the two adjacent atoms (Pb and the halide atoms) [40]. From a theoretical approach, it can be concluded that the smaller size of the atom can induce a large distance in synchronizing this approach with the addition of halide ions in MHPs ($APbX_3$), which results in tunning in the bandgap.

In comparison, the addition of chlorine and bromine atoms in the iodine MHPs induces a larger distance than the bromine atom with the Pb atom. It has an interesting result of tuning of the electronic bandgaps. Though there is no effect on the CBMs in the doping of Br or Cl atom in $MAPbI_3$, there is a huge impact on the VBM through the mixed contribution of the 5p orbitals of the iodine molecule with the 4p orbitals of the bromine molecule or the 3p orbital of the chlorine molecule, in the addition of Br or Cl in $MAPbI_3$, this result doping ratio of 0.67. In the near future, there is a huge chance of exploring Br–Cl combined mixed halide perovskites as the pure source for blue light emitters.

On the other hand, synthesizing the low dimensional perovskite (LDP) is essential in making blue light emitters. This type of synthesis includes quasi-2D, the formation of nanoplatelets, and the formation of nanocrystal perovskites. This synthesis type lies in the nanoscale crystals' quantum confinement effect near the exciton Bohr radius, which emits the blue emissive spectrum.

LDPs are preferred over the 3D mixed halide MHPs for their attractive advantages over 3D mixed halide MHPs. Mixed halide has operational instability due to phase segregation, whereas LDPs are free from this issue as there is no such use of mixed halide. LDPs include nanoscale crystal particles that exhibit large exciton binding energies compared with organic emitters. It also has larger bandgaps over the 3D bulk MHP crystals, which rely on the lattice strain regarding the quantum effect and the structural effects at the crystal surface [132]. In the case of large organic ligands like phenyl groups and aliphatic groups, segregation may be seen at the crystal's surface, forming a 2D crystal structure where a strong localization of exciton energy levels occurs in the LDPs, which explores a promising application for the near future of LEDs [133].

1.4.8 STUDY OF PHASE TRANSITION PROPERTY AND LOW-FIELD DIELECTRIC PROPERTY OF PEROVSKITE $CH_3NH_3PBI_3$ CRYSTALS

The limited edition of the source of fossil fuels will lead our near future to renewable energy sources. So further advancement for clean energy production is happening in full fledge mode. In advancing the PV solar cell section, perovskite takes attention by providing a low-cost solar cell with a moderate level (approximately exceeding 24%) of PCE. Another important property attracting interest is ferroelectric in bulk at room temperature. The existence of ferroelectricity is the main reason for converting light energy into electricity. The existence of the ferroelectricity in MAPI may be assumed to be the spontaneous and persistent distortion, rotation, and tilting of the octahedral PbI_6. Researchers can achieve a higher value of PCE, but we are facing challenges with the stability of the perovskite and the electrical performance of the PSC. An example of the performance of $CH_3NH_3PbI_3$ (MAPI) perovskite on its piezoelectric and electrostrictive properties depends on the strain induced when an electrical field is applied. This electrostrictive strain property between two PSCs greatly impacts the stability of perovskite or the degradation of perovskite under illumination [134]. There is a permanent dipolar property in the MAPI molecule, which prefers the polarized domain structure formation. There is an association property between electric property and elastic dipole properties on the highly rotationally efficient molecule like MA (CH_3NH_3). MA reorients itself and correlates with the tilt structure of octahedral PbI_6. This correlation results in the overall electric dipolar property between MA cation and octahedral PbI_6 anion of the perovskite [135–137].

With varying temperatures, an efficient change in the symmetry of crystals is due to the phase transition. In the MAPI molecule, the MA site is well known for its highly rotational behavior at room temperature, which shows a dynamic property of the MA molecule that results in the existence of ferroelectricity in the MAPI molecule, which makes an argument for the presence of ferroelectricity. In addition, the study of piezoelectric force microscopy of MAPI molecule claims and provides evidence for the existence of ferroelectricity [138, 139]. In contrast, some other studies claim no existence of the ferroelectric polarization property of MAPI molecule. The previous studies on MAPI provide different structural phase transitions, like the conversion from the orthorhombic structure to a tetragonal structure at the temperature of −113°C and from the orthorhombic structure to a paraelectric cubic phase over 54°C [140, 141], which results in a ferroelectric domain that is randomly oriented at room temperature. There was another controversy on the existence of ferroelectricity on the MAPI molecule through a study by Cordero et al. [141] who claimed that the reorientation dynamic through the combination of electric property and elastic dipolar property of the organic MA molecule is the main reason behind showing the ferroelectric property of MAPI molecule, has to compete with the distorting mode of PbI_6 molecule originating through the in-phase tilting mode of the PbI_6 molecule, which results in an antiferroelectric property of it, and ultimately the overall process results in the nullification of the ferroelectric behavior of the MAPI molecule [142].

In the case of dielectric spectroscopy's frequency range of 10 mHz/10 MHz, the dielectric permittivity [143, 144] was recorded by heating between the temperature range of −30°C and 110°C and parallelly dealing with different processes. The results can be explained in the three large regions. The first is the low-frequency range, where a little temperature change sharply changes the permittivity value. This result is an output of a combining process of the thermally activated method of the Maxwell–Wagner mechanism for the property of electrode polarization and dc-ionic conduction [145–147]. The second one is the moderate range of frequencies, which results in a moderate variation of permittivity with temperature. This result is associated with the charge accumulation process at the grain boundaries, which produces an extensive range of dielectric relaxation with the independence of the temperature [146, 147].

Moreover, another region is the high-frequency region, where it is observed that there is less variation of permittivity with temperature variation.

This process entirely depends on the response from the bulk of crystals [147]. It is also observed that dielectric losses increase on increasing the temperature, and this measured value will remain low over a temperature range of frequency above 10 kHz.

It is already known that the presence of organic cation (MA) causes the phase transition in the MAPI molecule as well as it is also responsible for showing the spontaneous polarization characteristic of this molecule by the disappearance of the centrosymmetric structure of this molecule; furthermore, it also indicates wilderness on its dynamic behavior at a preferably higher temperature [148]. It is also observed that the weak bonding of crystals and the weak cohesion force between two molecules based on the rotational background also become a reason for octahedral distortion as well as for ionic diffusion, according to Egger et al. [148]. In the MAPI molecule, along with the disorder dynamics of MA ion, PbI_6 also arrives with its displacive character and rearrangement property. At last, we can say that MAPI molecule undergoers go through an order–disorder phase transition, and this exists due to the suitable coupling between those properties of MA ion and PbI_6 ion [149–152]. Finally, correlating the cause behind the ionic polarization of MAPI with the mobile nature of MA molecule, including the displacive effect of the first-order phase transition.

There was a requirement of time for the dipoles and electrical charges after the polarization of the molecule, rearrangement and the alignment of the PbI_6 molecule, and the mobility of the MA molecule under the condition of applied AC electric field known as dielectric relaxations [153–155]. Permittivity has a high value at low frequencies, including the dielectric relaxations instead of the conductivity contribution. The analysis of complex permittivity and the dielectric modular approach gives a better result on the dependency on frequencies of MAPI molecules.

1.4.9 STUDY OF THERMALLY COUPLED AND UNCOUPLED ENERGY LEVEL OF DOUBLE PEROVSKITE MATERIAL LA_2MGTIO_6:ER^{3+} FOR ITS TEMPERATURE SENSING ABILITY

LMT in the DP family is important in introducing lanthanides. In research, it has been found that LMT: Er^{3+} can sense the temperature. The presence of thermally coupled energy levels and uncoupled energy levels ensures this molecule's behavior. The presence of those energy levels helps to have the highest sensitivity with the lowest uncertainty. The renowned thermally coupled energy level transitions are from $^2H_{11/2}$ energy level to $^4I_{15/2}$ energy level, and another transition is from $^4S_{3/2}$ energy level to $^4I_{15/2}$ energy level, where the absorption spectrum value of 592 cm^{-1} stays them apart. There is a suitable energy gap between those energy levels as they show a preferable energy gap to overlap each other efficiently; this energy gap value belongs to the range of 200 cm^{-1} absorption spectrum to 2000 cm^{-1} absorption spectrum [156]. Uncoupled levels include a phonon-assisted process, resulting in the temperature-sensing property of this DP containing lanthanide (LMT: Er^{3+}) [157–159]. The phonon-assisted process can be defined as the transition of energy levels like ($^2H_{11/2}$ + $^4S_{3/2}$→$^4F_{9/2,}$ $^4I_{15/2}$→$^4I_{13/2}$).

The emission spectrum analysis of LMT: Er^{3+} results in two opposite phenomena when the concentration of doping ions increases. One phenomenon is the increase in the emission intensity for all the energy bands when the concentration of Er^{3+} ion is increased in the LMT. This occurs when thermalization happens at 300 K, where $^2H_{11/2}$ and $^4S_{3/2}$ energy levels are concluded as one. In the 1–5% concentration range of Er^{3+}, it is observed at the local maxima of 3%, that the emission intensity of $^4S_{3/2}$ remains constant. On the other hand, an increase in emission intensity for the energy levels of $^4I_{11/2}$ and $^4F_{9/2}$ in the concentration range of 0.5% of Er^{3+} is observed. With the increasing concentration of dopant, the possibility of this process increases as it depends on the number of the Er^{3+} ion pair.

Mostly the intense emission can be seen in the transition of energy levels, like from $^4S_{3/2}$ →$^4I_{15/2}$ (I_3) at the concentration range of 5% Er^{3+} in LMT: Er^{3+}. By increasing the temperature through the thermalization process, mainly I_3 and other energy levels transition, like $^2H_{9/2}$→$^4I_{15/2}$ known as I_1, the transition from $^4S_{3/2}$→$^4I_{13/2}$ known as I_5, and the transition from $^4F_{3/2}$ →$^4I_{15/2}$ known as I_4, is depopulated strongly. Where I_3 works as the energy pump for the $^2H_{11/2}$→$^4I_{15/2}$ energy level transition. The maximum emission for the energy level I_2 can be observed at 248 K temperature. At this higher temperature, it is quenched quickly, whereas, at this temperature, the energy level transition decreases sharply for $^4I_{11/2}$ →$^4I_{15/2}$ transition. Before achieving this 248 K temperature, it results in stable energy transitions.

In the case of energy level coupled thermally, the $^4S_{3/2}$ energy level thermalization causes a strong population of the emission intensity for the energy transition of $^2H_{11/2}$ → $^4I_{15/2}$. But the emission intensity of the energy level transition comes from the energy level of $^4S_{3/2,}$ which is in the form of the energy level transition of I_3 and I_5. The steady quenching of the energy levels $^4S_{3/2} \rightarrow ^4I_{13/2}$ from $^4S_{3/2} \rightarrow ^4I_{15/2}$ results in the changes in thermometric parameters and a higher value of absolute sensitivity (S_a) for the energy transition of I_2/I_5. From the same energy level of $^4S_{3/2,}$ relative sensitivities (S_r) can be determined for the coupled energy level of I_3 and I_5. At the 148K temperature, the highest value of S_r can be obtained as 2.98% K^{-1} for the I_2/I_3 energy levels and 3.08% K^{-1} for I_2/I_5 energy levels. At the same temperature value of 148 K, the thermally uncoupled energy levels show a S_r value of 1.9% K^{-1} for the energy level pair I_2/I_1. This explains the temperature sensing property of the LMT: Er^{3+} perovskite containing lanthanides at a low-temperature range.

1.4.10 DOUBLE PEROVKITE IN THE CERAMIC FIELD AS BA_2MGWO_6

Due to the presence of Pb, widely used perovskite is not eco-friendly. If the tungstate molecule replaces the Pb molecule, it becomes eco-friendly perovskite. In the mode of advancement of the perovskite as DP, the tungstate DP introduced itself as ceramic in the form of BMW in 2004 for the planner resistance temperature detectors applications. In the recently published works on it as the synthesis of BMW, in addition to lanthanide molecules like Ce^{3+} by some traditional heating processes, which results in the form of BMW: $Ce^{3+,}$ which will be used in the advanced sintering technique, commonly known as the spark-plasma-sintering (SPS) method as scintillator based on ceramics [160]. SPS is also known as the densification process, where it works under high-pressure and high-temperature conditions for a conductive die, resulting in a high flow of electrical current.

Ceramics are highly preferred over any other materials for their attractive advantages. It is superior to other materials in the resistance value with the rapid temperature change. Also, it can work under high-temperature and high-pressure conditions along with rough conditions like a corrosive environment or a chemical environment. Ceramics also contain dispersion phenomena but significantly less exposure than other materials.

By the conventional solid-state method, BMW is synthesized along with different lanthanides like Eu^{3+} and Nd^{3+} doping or without lanthanide doping. This ceramic BMW works as a scintillator with or without doping of lanthanides. The X-ray excitation study for both types of molecules of BMW resulted in the visible region, proving it to work as a scintillator.

1.5 CHALLENGES OF PEROVSKITE SOLAR CELLS

With lots of potential, PSCs can be a prominent candidate to replace conventional energy harvesters. From 3.1% PCE in the early stage, PSC has come a long way where around 26% PCE has been achieved, and still, a lot more potential is there to adventure. One of the hindrances of industrialized PSC production is their production in the ambient atmosphere. Researchers have been trying to fabricate PSCs in open-air conditions [161]. There is a lot of work going on to find the best suitable material for HTM and ETM, which can achieve higher performance and better stability of devices. Being a toxic metal, Pb poses a grave environmental challenge. Getting rid of Pb completely, using other metals instead, and posing the same or better device performance along with better sustainability is a great challenge for researchers. There are a lot more ways to replace rigid Si-based solar cells completely. Although with many challenges, the industrialization of PSC is not far.

Nowadays, DPs are flourishing in their applications and the research, driving advancement and cost-effectiveness, which makes them suitable for both personal and commercial purposes. But in synthesizing it, researchers face different challenges of its disorder characteristic in the magnetic properties. Researchers only study the theoretical approach and predict the properties and the synthesis method for the DP containing 3d–4d/5d molecule, where we are also in the blind zone of its unknown properties like suitable doping material, pristine form of DP, different modification of organic cation and a well as the suitable divalent metal ions, revealing those properties may cause rapid progress in the research field of DP.

1.6 APPLICATION OF PEROVSKITE SOLAR CELLS

PSCs have various uses , and their applications are quite well known; it also has a wide range in the research field. The MHPs are widely used in the LED world [162], known as PeLEDs, and show a strong efficiency in their output results. It also has a dielectric property that increases PSCs' stability. MHPs have also been applied as a photocatalysts in fields like pollutant degradation, CO_2 reduction, or photocatalytic H_2 generation [163]. Vacancy defects introduced in semiconductors have also been proved to enhance photocatalytic activity [164].

Other than that, replacing perovskite consisting of Pb with perovskite consisting of tin also has huge applications. As the perovskite consists of tin ($CH_3NH_3SnCl_3$) also has a friendly impact on the environment, and in addition to that, it also has good optical and thermal properties, and that is why it is favored over the perovskite consisting of Pb [165,166]. Other perovskite molecules, like oxide-based perovskites, have shown usefulness in supercapacitors and new-age battery technology. $LaNiO_3$ and $LaMnO_3$ are being seen as a replacement for platinum in metal-air batteries due to their stable catalytic activity in the oxygen reduction reaction. Perovskite materials are also being used in photoelectrochemical water splitting [167]. Perovskite coated with hematite multilayer doped with TiO_2/Co is also used in the hydrogen production industry for its high potential [168].

But in the case of DP application, it also has various applications in different ways. As a scintillator, DP plays an essential role in the medical field. The ceramic background is a well-known material for producing both transparent and translucent ceramics. It is also used in the optical thermometer and as a dielectric material. This results in creating interest in the research field of DP.

1.7 COST-EFFECTIVENESS AND PRODUCTION OF INDUSTRIAL STANDARDS

PSCs have shown great promise as a low cost and highly efficient alternative to traditional silicon-based solar cells. However, their successful adoption in the industry requires addressing various factors related to cost-effectiveness and production scalability. Some of the key considerations are as follows [169, 170]:

1. Materials cost: PSCs are made using various materials, including perovskite precursors, hole and ETMs, substrates, and conductive layers. To achieve cost-effectiveness, optimizing the synthesis and sourcing of these materials is essential. Researchers are continuously working on developing less expensive and more readily

available materials without compromising the performance of solar cells.

2. Manufacturing process: PCSs' manufacturing process must be scalable and compatible with industrial production methods. Solution-based deposition techniques, such as spin-coating or inkjet printing, are promising for large-scale manufacturing due to their relative simplicity and potential for high-throughput production. The challenge is developing processes that consistently produce high-quality and uniform perovskite films on large substrates.
3. Stability and reliability: One of the primary concerns with PSCs is their stability and long-term reliability. Perovskite materials can be sensitive to moisture and environmental factors, leading to long-term degradation. Addressing stability issues is crucial to ensure that PSCs have a long operational lifetime and can withstand harsh environmental conditions, a significant requirement for industrial applications.
4. Efficiency and performance: Efficiency is a crucial driver for the cost-effectiveness of solar cells. Higher efficiency allows more electricity to be generated per unit area, reducing the overall cost per watt of electricity produced. Research efforts are focused on improving the power conversion efficiency of PSCs, making them more competitive with established solar technologies.
5. Standardization and certification: For widespread industrial adoption, there is a need for standardization and certification of PSCs. Developing standardized testing procedures and performance metrics will ensure consistency in evaluating and comparing different PSC technologies. Certification will also instill confidence among investors, consumers, and the industry, helping to drive further investments and market growth.
6. Large-scale manufacturing infrastructure: Manufacturers will need appropriate infrastructure and equipment to produce PSCs at an industrial scale. Existing solar cell manufacturing facilities may require modifications to accommodate the specific requirements of PSCs. Building a robust supply chain and infrastructure to support large-scale production is essential for cost-effectiveness and market penetration.

Overall, PSCs hold tremendous potential to transform the solar energy industry. Addressing cost-effectiveness, scalability, stability, and standardization challenges will be crucial for successfully integrating PSCs into the global energy market and realizing their full potential as a clean and sustainable energy solution.

1.8 CHARACTERISTIC RESULTS

Characteristic results in dielectric studies, electric studies, crystal characteristics, and optical studies of PSCs provide essential insights into their electronic and optoelectronic properties. PSCs have unique properties due to their perovskite crystal structure, and understanding these characteristics is critical for improving their efficiency and stability. Here are some characteristic results for each study [171]. Dielectric studies: Dielectric studies in PSCs focus on understanding the response of the material to an applied electric field. Some characteristic results include:

1. Dielectric constant (ε): The dielectric constant of the perovskite material determines its ability to store electrical energy when subjected to an electric field. Understanding the dielectric constant is crucial for optimizing the charge transport properties within the solar cell.
2. Frequency-dependent dielectric behavior: PSCs, known as dielectric dispersion, can exhibit frequency-dependent dielectric behavior. This dispersion can be studied across various frequencies to understand how the dielectric response changes with the applied electric field.
3. Dielectric loss: Dielectric loss in PSCs refers to the energy dissipated as heat due to dielectric polarization. Measuring the dielectric loss provides insights into the efficiency of charge carrier transport within the material.
4. Electric studies: Electric studies in PSCs aim to characterize their electrical properties. Some characteristic results include:

    a. Resistivity and conductivity: The resistivity and conductivity of the perovskite material are crucial for determining its ability to transport charge carriers. Lower resistivity and higher conductivity are desirable for efficient charge extraction.
    b. Carrier mobility: Carrier mobility measures the rate charge carriers can move through the material under an electric field. High carrier mobility is essential for efficient charge transport and collection.

c. Hall effect: The Hall effect can be used to determine the type and concentration of charge carriers in PSCs. It provides valuable information about the charge carrier dynamics within the material (Figure 1.3).

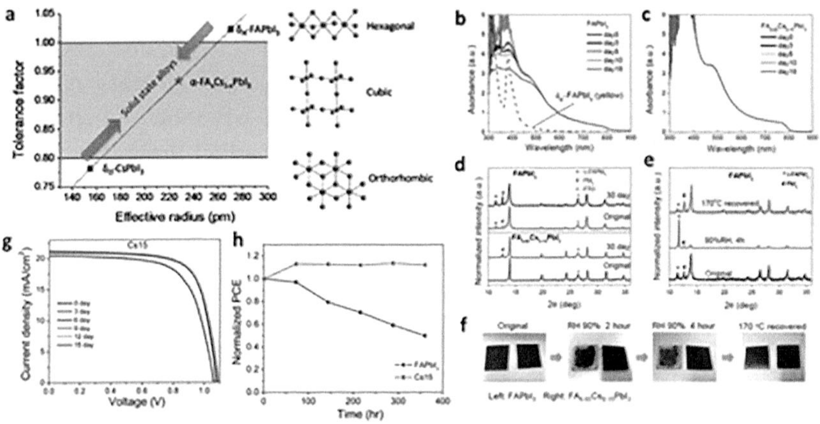

FIGURE 1.3 (a) Schematic illustration of the importance of the size of the A-site cation in determining the tolerance factor and structural properties of perovskite. Substitution of small amounts of FA with Cs on the A-site results in more structurally stable compounds. Changes in UV–vis spectra of (b) FAPbI3 and (c) FA0.85Cs0.15PbI3 over 18 days. (d) XRD patterns of original FAPbI3 and FA0.85Cs0.15PbI3 films and those films after 30 days of storage. (e) Change in XRD patterns of FAPbI3 thin film after its exposure to high humidity. (f) Photos of FAPbI3 and FA0.85Cs0.15PbI3 films under high-humidity conditions. (g) J–V curves of FA0.85Cs0.15PbI3 solar cells at 0–15 days of storage under 15% RH. (h) Normalized PCE of FAPbI3 and FA0.85Cs0.15PbI3 solar cells at different storage times.

Source: Reproduced with permission from Ref. [176]. Copyright 2016. American Chemical Society.

Crystal characteristics: Crystal characteristics are important to understand the structural properties of the perovskite material. Some characteristic results include:

1. X-ray diffraction (XRD) pattern: XRD can be used to identify the crystal structure and phase purity of the perovskite material. The XRD pattern reveals characteristic peaks corresponding to the atoms' arrangement in the crystal lattice.
2. Crystal symmetry: The crystal symmetry of the perovskite material is essential for understanding its electronic and optical properties.

Different crystal structures can affect charge carrier recombination and optoelectronic performance.

Optical studies: Optical studies in PSCs explore how the material interacts with light. Some characteristic results include:

1. UV spectrum: The UV spectrum of PSCs reveals the wavelengths of light that the material absorbs. It provides insights into the bandgap and optical properties of the perovskite material.
2. Photoluminescence spectrum: The photoluminescence spectrum shows the wavelengths of light emitted by the perovskite material when photons excite. It can indicate the presence of defects and trap states, which can impact the solar cell performance.
3. Refractive index: The refractive index of the perovskite material is crucial for designing antireflection coatings and understanding light management within the solar cell.
4. Transient absorption: Transient absorption spectroscopy is a versatile and powerful technique for exploring ultrafast processes in materials, helping to advance our understanding of fundamental physical and chemical phenomena. This is used to study charge carrier generation, transport, and recombination in solar cell materials, including PSCs. It allows the investigation of the relaxation pathways of photoexcited states, revealing information about energy transfer, charge carrier dynamics, and reaction mechanisms. It also provides insights into light-matter interactions in materials and enables the study of light-induced phenomena [173].

Understanding these characteristic results is essential for optimizing the efficiency, stability, and performance of PSCs and for further advancing their use in practical applications.

1.7 FUTURE ASPECT

As DP has not had a huge range in the research field like perovskite. Various advancements in DP are progressing daily alongside perovskite research. As perovskite has a wide range of applications compared to DP, now DP also expands slowly in various ways. This review discusses a small part of the advancements of perovskites and DP. The uses of MHPs in red and green PeLDEs have approximately reached their theoretical value regarding efficiency. But in the case of blue LEDs, it faces lag to reach an acceptable efficiency in the field of synthesis due to some characteristics like material instability, high proton energy [19, 20], and operational instability. The addition of quantum dots in solar cells is also an excellent approach for PV technology by using dielectric nanomaterials for quantum dots solar cells with increasing charge-carrying properties, which is based on the photoanode combining CdS–TiO_2 [174]. Future research is required to overcome those properties and reach an acceptable efficiency level by researching different processes like a thin film forming process and passivation of the defect. In this PV technology, defect engineering is important in solar cell stability and performance. Nowadays, we are facing trouble synthesizing blue PeLEDs, and we must search for a novel method for its synthesis process. We can also improve organic LEDs' luminescent stability and lifetime by researching and developing those processes. Further research is needed on the stability of perovskite through its dielectric property.

In the case of the temperature sensing ability of the DP, including lanthanide doping and without lanthanide doping, further research is required, as heating of energy levels that are coupled thermally shows the opposite behavior. We must increase our research by intending to increase the use of environmentally friendly DP in many fields, as we all are focused on the terms of 'no harm to the environment. The environmentally friendly DP can be synthesized using different molecules like tungstate in DP, fluorides DP, silicates DP, and sulfides DP; among them, tungstate in DP is mostly preferred. This proves further research is hugely needed to overcome that resistance from the perspective of developing clean energy technology.

1.8 CONCLUSIVE REMARK

With lots of potential, PSCs can be a prominent candidate to replace conventional energy harvesters. From 3.1% PCE in the early stage, PSC has come a long way where around 26% PCE has been achieved, and still a lot more potential is there to adventure. A lot of work is going on to find the best suitable material for HTM and ETM, which can achieve higher performance and better stability of devices. Being a toxic metal, Pb poses a grave environmental challenge. Getting rid of Pb completely, using other metals instead, posing the same or better device performance, and having better sustainability is a great challenge for researchers. There are a lot more ways to replace rigid Si-based solar cells completely. Although with many challenges, the industrialization of PSC is not far.

Exploring metal halides as PeLEDs is becoming a point of research interest, mainly synthesizing blue PeLEDs in dimensional engineering. Developing electrical conduction in the solar cell also takes a considerable place in studying its dielectric behavior and ferroelectricity in the perovskites. DP creates a revolution in the medical field by introducing itself as a scintillator. It also snatches the interest of scientists with its temperature-sensing ability. In the ceramic field, DP also improves the ability of ceramics to work in harsh conditions. Those improvements in the perovskite and DP research field are the main guides for further progress in clean energy development.

ACKNOWLEDGMENT

This work was supported by Science and Engineering Research Board [SERB] grants funded by the Department of Science and Technology [DST] Central, Government of India, through Teachers Associateship for Research Excellence [TAR/2018/000195].

KEYWORDS

- **fossil fuel**
- **perovskite solar cell**
- **electron transporting layer**
- **photovoltaic technology**

REFERENCES

1. Gratzel, M. Photoelectrochemical cells. *Nature* 2001, 414, 338–344.
2. Green, M.A., Emery, K., Hishikawa, Y., Warta, W., Dunlop, E.D. Solar cell efficiency tables. *Progress in Photovoltaics: Research and Applications* 2014, 22, 701–710.

3. Oregan, B., Gratzel, M. A low-cost, high-efficiency solar cell based on dye-sensitized colloidal TiO_2 films. *Nature* 1991, 353, 737–740.
4. Tang, C.W. Two-layer organic photovoltaic cell. *Applied Physics Letters* 1986, 48(2), 183.
5. Yu, G., Gao, J., Hummelen, J.C., Wudl, F., Heeger, A.J. Polymer photovoltaic cells: Enhanced efficiencies via a network of internal donor-acceptor heterojunctions. *Science* 1995, 270(5243), 1789–1791.
6. Bach, U., Lupo, D., Comte, P., Moser, J.E., Weissortel, F., Salbeck, J., Spreitzer, H., Gratzel, M. Solid-state dye-sensitized mesoporous TiO_2 solar cells with high photon-to-electron conversion efficiency. *Nature* 1998, 395, 583–585.
7. Kruger, J., Plass, R., Cevey, L., Piccirelli, M., Gratzel, M. High efficiency solid-state photovoltaic device due to inhibition of interface charge recombination. *Applied Physics Letters*, 2001, 79(13), 2085–2087.
8. Kojima, A., Teshima, K., Shirai, Y., Miyasaka, T. Organometal halide perovskites as visible-light sensitizers for photovoltaic cells. *Journal of the American Chemical Society* 2009, 131(17), 6050–6051.
9. Kim, H.S., Lee, C.R., Im, J.H., Lee, K.B., Moehl, T., Marchioro, A., Moon, S.J., Humphry-Baker, R., Yum, J.H., Moser, J.E., Gratzel, M., Park, N.G. Lead iodide perovskite sensitized all-solid-state submicron thin film mesoscopic solar cell with efficiency exceeding 9%. *Scientific Reports* 2012, 2, 591.
10. Lee, M.M., Teuscher, J., Miyasaka, T., Murakami, T.N., Snaith, H.J. Efficient hybrid solar cells based on meso-superstructured organometal halide perovskites. *Science* 2021, 338(6107), 643–647.
11. Kim, J.Y., Lee, J.-W., Jung, H.S., Shin, H., Park, N.-G., High-efficiency perovskite solar cells. *Chemical Reviews* 2020, 120(15), 7867–7918.
12. Liang, J., Han, X., Yang, J., Zhang, B., Fang, Q., Zhang, J., Ai, Q., Ogle, M.M., Terlier, T., Martí, A.A., Lou, J. Defect-engineering-enabled high-efficiency all-inorganic perovskite solar cells. *Advanced Materials* 2019, 31(51), 1903448.
13. Dou, L., Yang, Y., You, J., Hong, Z., Chang, W.-H., Li, G., Yang, Y. Solution-processed hybrid perovskite photodetectors with high detectivity. *Nature Communications* 2014, 5, 5404.
14. Zhu, Y., Liu, Y., Ai, Q., Gao, G., Yuan, L., Fang, Q., Tian, X., Zhang, X., Egap, E., Ajayan, P.M., Lou, J. In situ synthesis of lead-free halide perovskite–COF nanocomposites as photocatalysts for photoinduced polymerization in both organic and aqueous phases. *ACS Materials Letters* 2022, 4(3), 464–471.
15. Qin, C., Sandanayaka, A.S.D., Zhao, C., Matsushima, T., Zhang, D., Fujihara, T., Adachi, C. Stable room-temperature continuous-wave lasing in quasi-2D perovskite films. *Nature* 2020, 585, 53–57.
16. Lee, S., Kim, D.B., Yu, J.C., Jang, C.H., Park, J.H., Lee, B.R., Song, M.H. Versatile defect passivation methods for metal halide perovskite materials and their application to light-emitting devices. *Advanced Materials* 2019, 31(20), 1805244.
17. Hassan, Y., Park, J.H., Crawford, M.L., Sadhanala, A., Lee, J., Sadighian, J.C., Mosconi, E., Shivanna, R., Radicchi, E., Jeong, M., Yang, C., Choi, H., Park, S.H., Song, M.H., Angelis, F. De, Wong, C.Y., Friend, R.H., Lee, B.R., Snaith, H.J. Ligand-engineered bandgap stability in mixed-halide perovskite LEDs. *Nature* 2021, 591, 72–77.

18. Park, M.-H. 3D and 2D Metal halide-perovskites for blue light-emitting diodes. *Materials* 2022, 15(13), 4571.
19. Yeh, N., Chung, J.-P. High-brightness LEDs—Energy-efficient lighting sources and their potential in indoor plant cultivation. *Renewable and Sustainable Energy Reviews* 2009, 13(8), 2175–2180.
20. Yam, F.K., Hassan, Z. Innovative advances in LED technology. *Microelectronics Journal* 2005, 36(2), 129–137.
21. Ren, Z., Xiao, X., Ma, R., Lin, H., Wang, K., Sun, X.W., Choy, W.C.H. Hole transport bilayer structure for quasi-2D perovskite based blue light-emitting diodes with high brightness and good spectral stability. *Advanced Functional Materials* 2019, 29(43), 1905339.
22. Ren, Z., Yu, J., Qin, Z., Wang, J., Sun, J., Chan, C.C.S., Ding, S., Wang, K., Chen, R., Wong, K.S., Lu, X., Yin, W.-J., Choy, W.C.H. High-performance blue perovskite light-emitting diodes enabled by efficient energy transfer between coupled quasi-2D perovskite layers. *Advanced Materials* 2020, 33(1), 2005570.
23. Jin, Y., Wang, Z., Yuan, S., Wang, Q., Qin, C., Wang, K., Dong, C., Li, M., Liu, Y., Liao, L. Synergistic effect of dual ligands on stable blue quasi-2D Perovskite light-emitting diodes. *Advanced Functional Materials* 2019, 30(6), 1908339.
24. Wang, H., Zhang, X., Wu, Q., Cao, F., Yang, D., Shang, Y., Ning, Z., Zhang, W., Zheng, W., Yan, Y., Kershaw, S.V., Zhang, L., Rogach, A.L., Yang, X. Trifluoroacetate induced small-grained $CsPbBr_3$ perovskite films result in efficient and stable light-emitting devices. *Nature Communications* 2019, 10, 665.
25. Stefanska, D., Bondzior, B., Vu, T.H.Q., Grodzicki, M., Deren, P.J. Temperature sensitivity modulation through changing the vanadium concentration in a $La_2MgTiO_6:V^{5+}$, Cr^{3+} double perovskite optical thermometer. *Dalton Transactions* 2021, 50(28), 9851–9857.
26. Vu, T.H.Q., Bondzior, B., Stefanska, D., Deren, P.J. Influence of temperature on near-infrared luminescence, energy transfer mechanism and the temperature sensing ability of $La_2MgTiO_6:Nd^{3+}$ double perovskites. *Sensors and Actuators A: Physical* 2021, 317, 112453.
27. Bondzior, B., Stefanska, D., Vu, T.H.Q., Miniajluk-Gaweł, N., Deren, P.J. Red luminescence with controlled rise time in $La_2MgTiO_6:Eu^{3+}$. *Journal of Alloys and Compounds* 2021, 852, 157074.
28. Lu, L., Sun, M., Lu, Q., Wu, T., Huang, B. High energy X-ray radiation sensitive scintillating materials for medical imaging, cancer diagnosis and therapy. *Nano Energy* 2021, 79, 105437.
29. Xu, J., Shi, Y., Xie, J., Lei, F. Fabrication, microstructure, and luminescent properties of Ce^{3+}-doped $Lu_3Al_5O_{12}$ (Ce:LuAG) transparent ceramics by low-temperature vacuum sintering. *Journal of the American Ceramic Society* 2013, 96(6), 1930–1936.
30. Li, H.-L., Liu, X.-J., Huang, L.-P. Fabrication of transparent cerium-doped lutetium aluminum garnet (LuAG:Ce) ceramics by a solid-state reaction method. *Journal of the American Ceramic Society* 2005, 88(11), 3226–3228.
31. Zhu, W., Ma, W., Su, Y., Chen, Z., Chen, X., Ma, Y., Bai, L., Xiao, W., Liu, T., Zhu, H., Liu, X., Liu, H., Liu, X., Yang, Y.M. Low-dose real-time X-ray imaging with nontoxic double perovskite scintillators. *Light Science & Applications* 2020, 9, 1–10.

32. Arai, M., Mizoi, K., Fujimoto, Y., Koshimizu, M., Nakauchi, D., Yanagida, T., Asai, K. Tl_2NaYCl_6: A new self-activated scintillator possessing an elpasolite structure. *Journal of Materials Science: Materials in Electronics* 2021, 32, 7906–7912.
33. Roy, P., Waghmare, V., Maiti, T. Environmentally friendly $Ba_xSr_{2-x}TiFeO_6$double perovskite with enhanced thermopower for high temperature thermoelectric power generation. *RSC Advances* 2016, 6(60), 54636–54643.
34. Tuyet, D.T., Vu, T.H.Q., Bondzior, B., Deren, P.J., Velpula, R.T., Nguyen, H.P.T., Luu, A.T., Nguyen, Q.H.N., Nguyen, H.-D. Deep red fluoride dots-in-nanoparticles for high color quality micro white light-emitting diodes. *Optics Express* 2020, 28(18), 26189–26199.
35. Yu, Y., Xia, J., Liang, Y. Basic understanding of perovskite solar cells and passivation mechanism. *AIP Advances* 2022, 12, 055307.
36. Hariadi, M., Dzikri, I., Purnamaningsih, R.W., Poespawati, N.R. Fabrication of low cost perovskite solar cell under ambient conditions using only spin coating deposition method. *E3S Web of Conferences* 2018, 67, 01022.
37. Kulkarni, S.A., Baikie, T., Boix, P.P., Yantara, N., Mathews, N., Mhaisalkar, S. Band-gap tuning of lead halide perovskites using a sequential deposition process. *Journal of Materials Chemistry A* 2014, 2, 9221–9225.
38. Ke, W., Fang, G., Wang, J., Qin, P., Tao, H., Lei, H., et al. Perovskite solar cell with an efficient TiO_2 compact film. *ACS Applied Materials & Interfaces* 2014, 6, 15959–15965.
39. Cheng, X., Jing, L., Zhao, Y., Du, S., Ding, J., Zhou, T. Crystal orientation-dependent optoelectronic properties of $MAPbCl_3$ single crystals. *Journal of Materials Chemistry C* 2018, 6, 1579–1586.
40. Kumawat, N.K., Dey, A., Kumar, A., Gopinathan, S.P., Narasimhan, K.L., Kabra, D. Band gap tuning of $CH_3NH_3Pb(Br_{1-x}Cl_x)_3$ hybrid perovskite for blue electroluminescence. *ACS Applied Materials Interfaces* 2015, 7(24), 13119–13124.
41. Yang, S., Fu, W., Zhang, Z., Chen, H., Li, C.-Z. Recent advances in perovskite solar cells: Efficiency, stability and lead-free perovskite. *Journal of Materials Chemistry A* 2017, 5(23), 11462–11482.
42. Turkevych, I., Kazaoui, S., Ito, E., Urano, T., Yamada, K., Tomiyasu, H., Yamagishi, H., Kondo, M., Aramaki, S. Photovoltaic rudorffites: Lead-free silver bismuth halides alternative to hybrid lead halide perovskites. *ChemSusChem* 2017, 10(19), 3754–3759.
43. Giustino, F., Snaith, H.J. Toward lead-free Perovskite solar cells. *ACS Energy Letters* 2016, 1(6), 1233–1240.
44. Wu, T., Liu, X., Luo, X., Lin, X., Cui, D., Wang, Y., Segawa, H., Zhang, Y., Han, L. Lead-free tin perovskite solar cells. *Joule* 2021, 5(4), 863–886.
45. Jain, S.M., Edvinsson, T., Durrant, J.R. Green fabrication of stable lead-free bismuth based perovskite solar cells using a non-toxic solvent. *Communications Chemistry* 2019, 2, 91.
46. Park, B.W., Philippe, B., Zhang, X., Rensmo, H., Boschloo, G., Johansson, E.M. Bismuth based hybrid perovskites $A_3Bi_2I_9$ (A: methylammonium or cesium) for solar cell application. *Advanced Materials* 2015, 27(43), 6806–6813.
47. Hoye, R.L., Brandt, R.E., Osherov, A., Stevanović, V., Stranks, S.D., Wilson, M.W.B., Kim, H., Akey, A.J., Perkins, J.D., Kurchin, R.C., Poindexter, J.R., Wang, E.N.,

Bawendi, M.G., Bulovic, V., Buonassisi, T. Methylammonium bismuth iodide as a lead-free, stable hybrid organic–inorganic solar absorber. *Chemistry—A European Journal* 2015, 22(8), 2605–2610.
48. Singh, T., Miyasaka, T. Stabilizing the efficiency beyond 20% with a mixed cation perovskite solar cell fabricated in ambient air under controlled humidity. *Advanced Energy Materials* 2017, 8(3), 1700677.
49. Raghvendra, Kumar, R.R., Pandey, S., K. Performance evaluation and material parameter perspective of eco-friendly highly efficient $CsSnGeI_3$ perovskite solar cell. *Superlattices and Microstructures* 2019, 135, 106273.
50. Salim, T., Sun, S., Abe, Y., Krishna, A., Grimsdale, A.C., Lam, Y.M. Perovskite-based solar cells: Impact of morphology and device architecture on device performance. *Journal of Materials Chemistry A* 2015, 3(17), 8943–8969.
51. Huang, L., Hu, Z., Xu, J., Zhang, K., Zhang, J., Zhang, J., Zhu, Y. Efficient and stable planar perovskite solar cells with a non-hygroscopic small molecule oxidant doped hole transport layer. *Electrochimica Acta* 2016, 196(1), 328–336.
52. Krishna, A., Grimsdale, A.C. Hole transporting materials for mesoscopic perovskite solar cells—towards a rational design? *Journal of Materials Chemistry A* 2017, 5(32), 16446–16466.
53. Burschka, J., Pellet, N., Moon, S.-J., Humphry-Baker, R., Gao, P., Nazeeruddin, M.K., Grätzel, M. Sequential deposition as a route to high-performance perovskite-sensitized solar cells. *Nature* 2013, 499, 316–319.
54. Wang, S., Sina, M., Parikh, P., Uekert, T., Shahbazian, B., Devaraj, A., Meng, Y.S. Role of 4-tert-butylpyridine as a hole transport layer morphological controller in perovskite solar cells. *Nano Letters* 2016, 16(9), 5594–5600.
55. Li, M., Wang, Z.-K., Yang, Y.-G., Hu, Y., Feng, S.-L., Wang, J.-M., Gao, X.-Y., Liao, L.-S. Copper salts doped spiro-OMeTAD for high-performance perovskite solar cells. *Advanced Energy Materials* 2016, 6(21), 1601156.
56. Jiang, X., Yu, Z., Zhang, Y., Lai, J., Li, J., Gurzadyan, G.G., Yang, X., Sun, L. High-performance regular perovskite solar cells employing low-cost poly(ethylenedioxythiophene) as a hole-transporting material. *Scientific Reports* 2017, 7, 42564.
57. Luo, H., Lin, X., Hou, X., Pan, L., Huang, S., Chen, X. Efficient and air-stable planar perovskite solar cells formed on graphene-oxide-modified PEDOT:PSS hole transport layer. *Nano-Micro Letters* 2017, 9, 39.
58. Di Giacomo, F., Razza, S., Matteocci, F., D'Epifanio, A., Licoccia, S., Brown, T.M., Di Carlo, A. High efficiency $CH_3NH_3PbI_{(3-x)}Cl_x$ perovskite solar cells with poly(3-hexylthiophene) hole transport layer. *Journal of Power Sources* 2014, 251, 152–156.
59. Zhang, Y., Elawad, M., Yu, Z., Jiang, X., Lai, J., Sun, L. Enhanced performance of perovskite solar cells with P3HT hole-transporting materials via molecular p-type doping. *RSC Advances* 2016, 6(110), 108888–108895.
60. Jung, J.W., Park, J.-S., Han, I.K., Lee, Y., Park, C., Kwon, W., Park, M. Flexible and highly efficient perovskite solar cells with a large active area incorporating cobalt-doped poly(3-hexylthiophene) for enhanced open-circuit voltage. *Journal of Materials Chemistry A* 2017, 5(24), 12158–12167.

61. Lin, Q., Jiang, W., Zhang, S., Nagiri, R.C.R., Jin, H., Burn, P.L., Meredith, P. A triarylamine-based anode modifier for efficient organohalide perovskite solar cells. *ACS Applied Materials & Interfaces* 2017, 9(10), 9096–9101.
62. Zhou, Z., Li, X., Cai, M., Xie, F., Wu, Y., Lan, Z., Yang, X., Qiang, Y., Islam, A., Han, L. Stable inverted planar perovskite solar cells with low-temperature-processed hole-transport bilayer. *Advanced Energy Materials* 2017, 7(22), 1700763.
63. Jeon, N.J., Lee, J., Noh, J.H., Nazeeruddin, M.K., Grätzel, M., Seok, S.I. Efficient inorganic–organic hybrid perovskite solar cells based on pyrene arylamine derivatives as hole-transporting materials. *Journal of the American Chemical Society* 2013, 51(135), 19087–19090.
64. Yang, L., Cai, F., Yan, Y., Li, J., Liu, D., Pearson, A.J., Wang, T. Conjugated small molecule for efficient hole transport in high-performance p-i-n type perovskite solar cells. *Advanced Functional Materials* 2017, 27(31), 1702613.
65. Chen, Z., Li, H., Zheng, X., Zhang, Q., Li, Z., Hao, Y., Fang, G. Low-cost carbazole-based hole transport material for highly efficient perovskite solar cells. *ChemSusChem* 2017, 10(15), 3111–3117.
66. Malinauskas, T., Saliba, M., Matsui, T., Daskeviciene, M., Urnikaite, S., Gratia, P., Send, R.,Wonneberger, H., Bruder, I., Graetzel, M., Getautis, V., Nazeeruddin, M.K., Branched methoxydiphenylamine-substituted fluorene derivatives as hole transporting materials for high-performance perovskite solar cells. *Energy & Environmental Science* 2016, 9(5), 1681–1686.
67. Gratia, P., Magomedov, A., Malinauskas, T., Daskeviciene, M., Abate, A., Ahmad, S., Gratzel, M., Getautis, V., Nazeeruddin, M.K. A methoxydiphenylamine-substituted carbazole twin derivative: An efficient hole-transporting material for perovskite solar cells. *Angewandte Chemie (International Edition in English)* 2015, 54(39), 11409–11413.
68. Irwin, M.D., Buchholz, D.B., Hains, A.W., Chang, R.P.H., Marks, T.J. p-type semiconducting nickel oxide as an efficiency-enhancing anode interfacial layer in polymer bulk-heterojunction solar cells. *Proceedings of the National Academy of Sciences* 2008, 8(105), 2783–2787.
69. Cui, J., Meng, F., Zhang, H., Cao, K., Yuan, H., Cheng, Y., Huang, F., Wang, M. $CH_3NH_3PbI_3$-based planar solar cells with magnetron-sputtered nickel oxide. *ACS Applied Materials & Interfaces* 2014, 6(24), 22862–22870.
70. Li, M.-H., Shen, P.-S., Wang, K.-C., Guo, T.-F., Chen, P. Inorganic p-type contact materials for perovskite-based solar cells. *Journal of Materials Chemistry A* 2015, 3(17), 9011–9019.
71. Zhao, X., Zhang, W., Feng, X., Guo, X., Lu, C., Li, X., Fang, J. Photoconductive NiO_x hole transport layer for efficient perovskite solar cells. *Chemical Engineering Journal* 2022, 3(435), 135140.
72. Kim, J.H., Liang, P.-W., Williams, S.T., Cho, N., Chueh, C.-C., Glaz, M.S., Ginger, D.S., Jen, A.K.Y. High-performance and environmentally stable planar heterojunction perovskite solar cells based on a solution-processed copper-doped nickel oxide hole-transporting layer. *Advanced Materials* 2014, 27(4), 695–701.
73. Makming, P., Homnan, S., Ruankham, P., Wongratanaphisan, D., Chimupala, Y., Goubard, F., Adjaoud, A., Intaniwet, A. Utilization of zinc-doped nickel oxide hole

transporting materials to improve efficiency and stability of perovskite solar cell. *Suan Sunandha Science and Technology Journal* 2022, 9(2), 71–78.
74. Cao, K., Zuo, Z., Cui, J., Shen, Y., Moehl, T., Zakeeruddin, S.M., Grätzel, M., Wang, M. Efficient screen printed perovskite solar cells based on mesoscopic $TiO_2/Al_2O_3/NiO$/carbon architecture. *Nano Energy* 2015, 17, 171–179.
75. Chen, W.-Y., Deng, L.-L., Dai, S.-M., Wang, X., Tian, C.-B., Zhan, X.-X., Xie, S.-Y., Huang, R.-B., Zheng, L.-S. Low-cost solution-processed copper iodide as an alternative to PEDOT:PSS hole transport layer for efficient and stable inverted planar heterojunction perovskite solar cells. *Journal of Materials Chemistry A* 2015, 3(38), 19353–19359.
76. Qin, P.-L., He, Q., Chen, C., Zheng, X.-L., Yang, G., Tao, H., Xiong, L.-B., Xiong, L.,Li, G., Fang, G.-J. High-performance rigid and flexible perovskite solar cells with low-temperature solution-processable binary metal oxide hole-transporting materials. *Solar RRL* 2017, 1(8), 1700058.
77. Zhang, H., Wang, H., Chen, W., Jen, A.K.Y. $CuGaO_2$: A promising inorganic hole-transporting material for highly efficient and stable perovskite solar cells. *Advanced Materials* 2016, 29(8), 1604984.
78. Zhang, F., Yang, X., Cheng, M., Wang, W., Sun, L. Boosting the efficiency and the stability of low cost perovskite solar cells by using CuPc nanorods as hole transport material and carbon as counter electrode. *Nano Energy* 2016, 20, 108–116.
79. Wang, J.-M., Wang, Z.-K., Li, M., Zhang, C.-C., Jiang, L.-L., Hu, K.-H., Ye, Q.-Q., Liao, L.-S. Doped Copper phthalocyanine via an aqueous solution process for normal and inverted perovskite solar cells. *Advanced Energy Materials* 2017, 8(2), 1701688.
80. Ito, S., Tanaka, S., Vahlman, H., Nishino, H., Manabe, K., Lund, P. Carbon-double-bond-free printed solar cells from $TiO_2/CH_3NH_3PbI_3$/CuSCN/Au: Structural control and photoaging effects. *ChemPhysChem* 2014, 15(6), 1194–1200.
81. Yang, B., Chen, Y., Cui, Y., Liu, D., Xu, B., Hou, J. Over 100-nm-thick MoO_x films with superior hole collection and transport properties for organic solar cells. *Advanced Energy Materials* 2018, 8(25), 1800698.
82. Lou, X.W., Zeng, H.C. Complex α-MoO_3 nanostructures with external bonding capacity for self-assembly. *Journal of American Chemical Society* 2003, 9(125), 2697–2704
83. Pan, W., Tian, R., Jin, H., Guo, Y., Zhang, L., Wu, X., Zhang, L., Han, Z., Liu, G., Li, J., Rao, G., Wang, H., Chu, W. Structure, optical, and catalytic properties of novel hexagonal metastable h-MoO_3 nano-and microrods synthesized with modified liquid-phase processes. *Chemistry of Materials* 2010, 22(22), 6202–6208.
84. Kyaw, A.K.K., Sun, X.W., Jiang, C.Y., Lo, G.Q., Zhao, D.W., Kwong, D.L. An inverted organic solar cell employing a sol-gel derived ZnO electron selective layer and thermal evaporated MoO_3 hole selective layer. *Applied Physics Letters* 2008, 93, 221107.
85. Cai, P., Ren, P., Huang, X., Zhang, X., Zhan, T., Xiong, J., Xue, X., Wang, Z., Zhang, J., Chen, J. An ultraviolet-deposited MoO_3 film as anode interlayer for high-performance polymer solar cells. *Advanced Materials Interfaces* 2020, 7(6), 1901912.
86. Chang, F.-K., Huang, Y.-C., Jeng, J.-S., Chen, J.-S. Band offset of vanadium-doped molybdenum oxide hole transport layer in organic photovoltaics. *Solid-State Electronics* 2016, 122, 18–22.

87. Bai, Y., Zhao, C., Guo, Q., Zhang, J., Hu, S., Liu, J., Hayat, T., Alsaedi, A., Tan, Z. Enhancing the electron blocking ability of n-type MoO_3 by doping with p-type NiO for efficient nonfullerene polymer solar cells. *Organic Electronics* 2019, 68, 168–175.
88. Li, Z., Guo, W., Liu, C., Zhang, X., Li, S., Guo, J., Zhang, L. Impedance investigation of the highly efficient polymer solar cells with composite $CuBr_2$ /MoO_3 hole transport layer. *Physical Chemistry Chemical Physics* 2017, 19(31), 20839–20846.
89. Brabec, C.J. Organic photovoltaics: Technology and market. *Solar Energy Materials and Solar Cells* 2004, 83(2–3), 273–292.
90. Lee, Y.H., Kim, D.H., Kim, T.W. Enhanced power conversion efficiency of organic photovoltaic devices due to the surface plasmonic resonance effect generated utilizing Au-WO_3 nanocomposites. *Organic Electronics* 2017, 45, 256–262.
91. Shen, W., Tang, J., Wang, D., Yang, R., Chen, W., Bao, X., Wang, Y., Jiao, J., Wang, Y., Huang, Z., Huang, L., Liu, J., Wang, W., Wu, P., Belfiore, L. Enhanced efficiency of polymer solar cells by structure-differentiated silver nano-dopants in solution-processed tungsten oxide layer. *Materials Science and Engineering: B* 2016, 206, 61–68.
92. Kung, P.-K., Li, M.-H., Lin, P.-Y., Chiang, Y.-H., Chan, C.-R., Guo, T.-F., Chen, P. A review of inorganic hole transport materials for perovskite solar cells. *Advanced Materials Interfaces* 2018, 5(22), 1800882.
93. Singh, T., Singh, J., Miyasaka, T. Role of metal oxide electron-transport layer modificationon the stability of high performance perovskite solar cells. *ChemSusChem* 2016, 9(18), 2559–2566.
94. Lian, J., Lu, B., Niu, F., Zeng, P., Zhan, X. Electron-transport materials in perovskite solar cells. *Small Methods* 2018, 2(10), 1800082.
95. Said, A.A., Xie, J., Zhang, Q. Recent progress in organic electron transport materials in inveted perovskite solar cells. *Small* 2019, 15(27), 1900854.
96. Mo, S.D., Ching, W.Y. Electronic and optical properties of three phases of titanium dioxide: Rutile, anatase, and brookite. *Physical Review B* 1995, 51, 13023.
97. Xie, F., Zhu, J., Li, Y., Shen, D., Abate, A., Wei, M. TiO_2-B as an electron transporting material for highly efficient perovskite solar cells. *Journal of Power Sources* 2019, 415, 8–14.
98. Kim, M., Kim, G.-H., Lee, T.K., Choi, I.W., Choi, H.W., Jo, Y., Yoon, Y.J., Kim, J.W., Lee, J., Huh, D., Lee, H., Kwak, S.K., Kim, J.Y., Kim, D.S. Methylammonium chloride induces intermediate phase stabilization for efficient perovskite solar cells. *Joule* 2019, 3(9), 2179–2192.
99. Wang, Y., Wan, J., Ding, J., Hu, J., Wang, D. A rutile TiO_2 electron transport layer for the enhancement of charge collection for efficient perovskite solar cells. *Angewandte Chemie International Edition* 2019, 58(28), 9414–9418.
100. Lu, H., Zhong, J., Ji, C., Zhao, J., Li, D., Zhao, R., Jiang, Y., Fang, S., Liang, T., Li, H., Li. C.M. Fabricating an optimal rutile TiO_2 electron transport layer by delicately tuning $TiCl_4$ precursor solution for high performance perovskite solar cells. *Nano Energy* 2020, 68, 104336.
101. Ke, W., Fang, G., Liu, Q., Xiong, L., Qin, P., Tao, H., Wang, J., Lei, H., Li, B., Wan, J., Yang, G., Yan, Y. Low-temperature solution-processes tin oxide as an alternative

electron transporting layer for efficient perovskite solar cells. *Journal of American Chemical Society* 2015, 137, 21, 6730–6733.

102. Ke, W., Zhao, D., Cimaroli, A.J., Grice, C.R., Qin, P., Liu, Q., Xiong, L., Yan, Y., Fang, G. Effects of annealing temperature of tin oxide electron selective layers on the performance of perovskite solar cells. *Journal of Materials Chemistry A* 2015, 3(47), 24163–24168.
103. Dong, Q., Shi, Y., Wang, K., Li, Y., Wang, S., Zhang, H., Xing, Y., Ma, T. Insight into perovskite solar cells based in SnO_2 compact electron-selective layer. *The Journal of Physical Chemistry C* 2015, 119, 19, 10212–10217.
104. Xiong, L., Guo, Y., Wen, J., Liu, H., Yang, G., Qin, P., Fang, G. Review on the application of SnO_2 in perovskite solar cells. *Advanced Functional Materials* 2018, 28(35), 1802757.
105. Jiang, Q., Zhang, X., You, J. SnO_2: A wonderful electron transport layer for perovskite solar cells. *Small* 2018, 14(31), 1801154.
106. Dong, Q., Shi, Y., Zhang, C., Wu, Y., Wang, L. Energetically favoured formation of SnO_2 nanocrystals as electron transfer layer in perovskite solar cells with high efficiency exceeding 19%. *Nano Energy* 2017, 40, 336–344.
107. Bai, Y., Fang, Y., Deng, Y., Wang, Q., Zhao, J., Zheng, X., Zhang, Y. Low-temperature solution processed Sb:SnO_2 nanocrystals for efficient planar perovskite solar cells. *ChemSusChem* 2016, 9(18), 2686–2691.
108. Park, M., Kim, J.-Y., Son, H.J., Lee, C.-H., Jang, S.S., Ko, M.J. Low-temperature solution-processed Li-doped SnO_2 as an effective electron transporting layer for high-performance flexible and wearable perovskite solar cells. *Nano Energy* 2016, 26, 208–215.
109. Qiu, L., Xing, K., Zhang, J., Yang, Y., Cao, W., Zhou, X., Zhu, K., Xia, D., Fan, R. Two-dimensional metal-organic frameworks-based grain termination strategy enables high-efficiency perovskite photovoltaics with enhanced moisture and thermal stability. *Advanced Functional Materials* 2021, 31(17), 2010368.
110. Anaraki, E.H., Kermanpur, A., Mayer, M.T., Steier, L., Ahmed, T., Turren-Cruz, S.-H., Seo, J., Luo, J., Zakeeruddin, S.M., Tress, W.R., Edvinsson, T., Grätzel, M., Hagfeldt, A., Correa-Baena, J.-P. Low-temperature Nb-doped SnO_2 electron-selective contact yields over 20% efficiency in planar perovskite solar cells. *ACS Energy Letters* 2018, 3(4), 773–778.
111. Liu, Q., Zhang, X., Li, C., Lu, H., Weng, Z., Pan, Y., Chen, W., Hang, X.-C., Sun, Z., Zhan, Y. Effect of tantalum doping on SnO_2 electron transport layer via low temperature process for perovskite solar cells. *Applied Physics Letters* 2019, 115, 14, 143903.
112. Liu, X., Bu, T., Li, J., He, J., Li, T., Zhang, J., Li, W., Ku, Z., Peng, Y., Huang, F., Cheng, Y.B., Zhong, J. Stacking n-type layers: Effective route towards stable, efficient and hysteresis-free planar perovskite solar cells. *Nano Energy* 2018, 44, 34–42.
113. Wang, D., Wu, C., Luo, W., Guo, X., Qu, B., Xiao, L., Chen, Z. ZnO/SnO_2 double electron transport layer guides improved open circuit voltage for highly efficient $CH_3NH_3PbI_3$-based planar perovskite solar cells. *ACS Applied Energy Materials* 2018, 1(5), 2215–2221.

114. Zheng, D., Wang, G., Huang, W., Wang, B., Ke, W., Logsdon, J.L., Wang, H., Wang, Z., Zhu, W., Yu, J., Wasielewski, M.R., Kanatzidis, M.G., Marks, T.J., Facchetti, A. Combustion synthesized zinc oxide electron-transport layers for efficient and stable perovskite solar cells. *Advanced Functional Materials* 2019, 29(16), 1900265.
115. Liu, D., Kelly, T.L. Perovskite solar cells with a planar heterojunction structure prepared using room-temperature solution processing techniques. *Nature Photonics* 8, 133–138, 2014.
116. Luo, J., Wang, Y., Zhang, Q. Progress in perovskite solar cells bsed on ZnO nanostructure. *Solar Energy* 2018, 163, 289–306.
117. Zhang, P., Wu, J., Zhang, T., Wang, Y., Liu, D., Chen, H., Ji, L., Liu, C., Ahmed, W., Chen, Z.D., Li, S. Perovskite solar cells with ZnO electron-transporting materials. *Advanced Materials* 2017, 30(3), 1703737.
118. Yang, J., Siempelkamp, B.D., Mosconi, E., Angelis, F. De, Kelly, T.L. Origin of the thermal instability in $CH_3NH_3PbI_3$ thin flims deposited on ZnO. *Chemistry of Materials* 2015, 27(12), 4229–4236.
119. Shin, S.S., Yeom, E.J., Yang, W.S., Hur, S., Kim, M.G., Im, J., Seo, J., Noh, J.H., Seok, S. Colloidally prepared La-doped $BaSnO_3$ electrodes for efficient, photostable perovskite solar cells. *Science* 2017, 356(6334), 167–171.
120. Yuan, S., Wang, Z.-K., Xiao, L-X.., Zhang, C.-F., Yang, S.-Y., Chen, B.-B., Ge, H.-T., Tian, Q.-S., Jin, Y., Liao, L.-S. Optimization of low-dimensional components of quasi-2D perovskite films for deep-blue light-emitting diodes. *Advanced Materials* 2019, 31(44), 1904319.
121. Song, L., Huang, L., Liu, Y., Hu, Y., Guo, X., Chang, Y., Geng, C., Xu, S., Zhang, Z., Zhang, Y., Luan, N. Efficient and stable blue perovskite light-emitting devices based on inorganic Cs_4PbBr_6 spaced low-dimensional $CsPbBr_3$ through synergistic control of amino alcohols and polymer additives. *ACS Applied Materials Interfaces* 2021, 13(28), 33199–33208.
122. Jamaludin, N.F., Yantara, N., Febriansyah, B., Tay, Y.B., Muhammad, B.T., Laxmi, S., Lim, S.S., Sum, T.C., Mhaisalkar, S., Mathews, N. Additives in halide perovskite for blue-light-emitting diodes: Passivating agents or crystallization modulators? *ACS Energy Letters* 2021, 6(12), 4265–4272.
123. Wu, C., Wu, T., Yang, Y., McLeod, J.A., Wang, Y., Zou, Y., Zhai, T., Li, J., Ban, M., Song, T., Gao, X., Duhm, S., Sirringhaus, H., Sun, B. Alternative type two-dimensional–three-dimensional lead halide perovskite with inorganic sodium ions as a spacer for high-performance light-emitting diodes. *ACS Nano* 2019, 13(2), 1645–1654.
124. Jang, C.H., Harit, A.K., Lee, S., Kim, S.H., Jeong, J.-E., Park, J.H., Jung, E.D., Ha, J.M., Kwak, S.K., Woo, H.Y., Song, M.H. Sky-blue-emissive perovskite light-emitting diodes: Crystal Growth and interfacial control using conjugated polyelectrolytes as a hole-transporting layer, *ACS Nano* 2020, 14(10), 13246–13255.
125. Liu, Y., Zhang, L., Chen, S., Liu, C., Li, Y., Wu, J., Wang, D., Jiang, Z., Li, Y., Li, Y., Wang, X., Xu, B. Water-soluble conjugated polyelectrolyte hole transporting layer for efficient sky-blue perovskite light-emitting diodes. *Small* 2021, 17(37), 2101477.
126. Gangishetty, M.K., Hou, S., Quan, Q., Congreve, D.N. Reducing architecture limitations for efficient blue perovskite light-emitting diodes. *Advanced Materials* 2018, 30(20), 1706226.

127. Buin, A., Comin, R., Xu, J., Ip, A.H., Sargent, E.H. Halide-dependent electronic structure of organolead perovskite materials. *Chemistry of Materials* 2015, 27(12), 4405–4412.
128. Tao, S., Schmidt, I., Brocks, G., Jiang, J., Tranca, I., Meerholz, K., Olthof, S. Absolute energy level positions in tin- and lead-based halide perovskites. *Nature Communications* 2019, 10, 2560.
129. Jiang, M., Deng, N., Wang, L., Xie, H., Qiu, Y. The structural, electronic, and optical properties of organic-inorganic mixed halide perovskites $CH_3NH_3Pb(I_{1-y}X_y)_3$(X = Cl, Br). *Chinese Physics B* 2018, 27(6), 067102.
130. Kitazawa, N., Watanabe, Y., Nakamura, Y. Optical properties of $CH_3NH_3PbX_3$(X = halogen) and their mixed-halide crystals. *Journal of Materials Science* 2002, 37, 3585–3587.
131. Wang, S., Bi, C., Yuan, J., Zhang, L., Tian, J. Original core-shell structure of cubic $CsPbBr_3$@Amorphous $CsPbBr_x$ perovskite quantum dots with a high blue photoluminescence quantum yield of over 80%. *ACS Energy Letters* 2018, 3(1), 245–251.
132. Gong, X., Voznyy, O., Jain, A., Liu, W., Sabatini, R., Piontkowski, Z., Walters, G., Bappi, G., Nokhrin, S., Bushuyev, O., Yuan, M., Comin, R., McCamant, D., Kelly, S.O., Sargent, E.H. Electron-phonon interaction in efficient perovskite blue emitters. *Nature Materials* 2018, 17, 550–556.
133. Chen, B., Li, T., Dong, Q., Mosconi, E., Song, J., Chen, Z., Deng, Y., Liu, Y., Ducharme, S., Gruverman, A., Angelis, F. De, Huang, J. Large electrostrictive response in lead halide perovskites. *Nature Materials* 2018, 17, 1020–1026.
134. Frost, J.M., Butler, K.T., Brivio, F., Hendon, C.H., van Schilfgaarde, M., Walsh, A. Atomistic origins of high-performance in hybrid halide perovskite solar cells. *Nano Letters* 2014, 14(5), 2584–2590.
135. Li, W., Wang, Z., Deschler, F., Gao, S., Friend, R.H., Cheetham, A.K. Chemically diverse and multifunctional hybrid organic–inorganic perovskites. *Nature Reviews Materials* 2017, 2, 16099.
136. Frost, J.M., Butler, K.T., Walsh, A. Molecular ferroelectric contributions to anomalous hysteresis in hybrid perovskite solar cells. *APL Materials* 2014, 2(8), 081506.
137. Chen, B., Shi, J., Zheng, X., Zhou, Y., Zhu, K., Priya, S. Ferroelectric solar cells based on inorganic–organic hybrid perovskites. *Journal of Materials Chemistry A* 2015, 3(15), 7699–7705.
138. Kutes, Y., Ye, L., Zhou, Y., Pang, S., Huey, B.D., Padture, N.P. Direct observation of ferroelectric domains in solution-processed $CH_3NH_3PbI_3$ perovskite thin films. *Journal of Physical Chemistry Letters* 2014, 5(19), 3335–3339.
139. Coll, M., Gomez, A., Mas-Marzá, E., Almora, O., Garcia-Belmonte, G., Campoy-Quiles, M., Bisquert, J. Polarization switching and light-enhanced piezoelectricity in lead halide perovskites, *Journal of Physical Chemistry Letters* 2015, 6(8), 1408–1413.
140. Fan, Z., Xiao, J., Sun, K., Chen, L., Hu, Y., Ouyang, J., Ong, K.P., Zeng, K., Wang, J. Ferroelectricity of $CH_3NH_3PbI_3$ perovskite. *Journal of Physical Chemistry Letters* 2015, 6(7), 1155–1161.
141. Cordero, F., Craciun, F., Trequattrini, F., Imperatori, P., Paoletti, A.M., Pennesi, G., Competition between polar and antiferrodistortive modes and correlated dynamics

of the methylammonium molecules in MAPbI3 from anelastic and dielectric measurements. *Journal of Physical Chemistry Letters* 2018, 9(15), 4401–4406.
142. Stoumpos, C.C., Malliakas, C.D., Kanatzidis, M.G. Semiconducting tin and lead iodide perovskites with organic cations: phase transitions, high mobilities, and near-infrared photoluminescent properties. *Inorganic Chemistry* 2013, 52(15), 9019–9038.
143. Baikie, T., Fang, Y., Kadro, J.M., Schreyer, M., Wei, F., Mhaisalkar, S.G., Graetzel, M., White, T.J. Synthesis and crystal chemistry of the hybrid perovskite $(CH_3NH_3)PbI_3$ for solid-state sensitised solar cell applications. *Journal of Materials Chemistry A* 2013, 1(18), 5628–5641.
144. Kremer, F., Schonhals, A. *Broadband Dielectric Spectroscopy*. Springer: Berlin/Heidelberg, Germany, 2003, pp. 58–98.
145. Ciomaga, C.E., Guzu, A., Airimioaei, M., Curecheriu, L.P., Lukacs, V.A., Avadanei, O.G., Stoian, G., Grigoras, M., Lupu, N., Asandulesa, M., Mitoseriu, L. Comparative study of magnetoelectric $BaTiO_3$–$Co_{0.8}Zn_{0.2}Fe_2O_4$ Bi-tunable ceramics sintered by spark plasma sintering and classical method. *Ceramics International* 2019, *45*, 18(A), 24168–24175.
146. Ciomaga, C.E., Neagu, A.M., Pop, M.V., Airimioaei, M., Tascu, S., Schileo, G., Galassi, C., Mitoseriu, L. Ferroelectric and dielectric properties of ferrite-ferroelectric ceramic composites. *Journal of Applied Physics* 2013, 113(7), 074103.
147. You, Y.-M., Liao, W.-Q., Zhao, D., Ye, H.-Y., Zhang, Y., Zhou, Q., Niu, X., Wang, J., Li, P.-F., Fu, D.-W., Wang, Z., Gao, S., Yang, K., Liu, J.-M., Li, J., Yan, Y., Xiong, R.-G. An organic-inorganic perovskite ferroelectric with large piezoelectric response. *Science* 2017, 357(6348), 306–309.
148. Egger, D.A., Rappe, A.M., Kronik, L. Hybrid organic-inorganic perovskites on the move. *Accounts of Chemical Research* 2016, 49(3), 573–581.
149. Mashiyama, H., Kawamura, Y., Magome, E., Kubota, Y. Displacive character of the cubic-tetragonal transition in $CH_3NH_3PbX_3$. *Journal of the Korean Physical Society* 2003, 42, 1026–1029.
150. Swainson, I.P., Stock, C., Parker, S.F., Van Eijck, L., Russina, M., Taylor, J.W. From soft harmonic phonons to fast relaxational dynamics in $CH_3NH_3PbBr_3$. *Physical Review B Covering Condensed Matter and Materials Physics* 2015, 92(10), 100303.
151. Brivio, F., Frost, J.M., Skelton, J.M., Jackson, A.J., Weber, O.J., Weller, M.T., Goñi, A.R., Leguy, A.M.A., Barnes, P.R.F., Walsh, A. Lattice dynamics and vibrational spectra of the orthorhombic, tetragonal, and cubic phases of methylammonium lead iodide. *Physical Review B Covering Condensed Matter and Materials Physics*, 2015, 92(14), 144308.
152. Whitfield, P.S., Herron, N., Guise, W.E., Page, K., Cheng, Y.Q., Milas, I., Crawford, M.K. Structures, phase transitions and tricritical behavior of the hybrid perovskite methyl ammonium lead iodide. *Scientific Reports* 2016, 6, 35685.
153. Knop, O., Wasylishen, R.E., White, M.A., Cameron, T.S., Van Voort, M.J.M. Alkylammonium lead halides. Part 2. $CH_3NH_3PbX_3$(X = C1, Br, I) perovskites: Cuboctahedral halide cages with isotropic cation reorientation. *Canadian Journal of Chemistry* 1990, 68, 412–422.
154. Onoda-Yamamuro, N., Matsuo, T., Suga, H. Calorimetric and IR spectroscopic studies of phase transitions in methylammonium trihalogenplumbates (II). *Journal of Physics and Chemistry of Solids* 1990, 51(12), 1383–1395.

155. Wade, S.A., Collins, S.F., Baxter, G.W. Fluorescence intensity ratio technique for optical fiber point temperature sensing. *Journal of Applied Physics* 2003, 94(8), 4743–4756.
156. Brandão-Silva, A.C., Gomes, M.A., Novais, S.M.V., Macedo, Z.S., Avila, J.F.M., Rodrigues, J.J., Alencar, M.A.R.C. Size influence on temperature sensing of erbium-doped yttrium oxide nanocrystals exploiting thermally coupled and uncoupled levels' pairs. *Journal of Alloys and Compounds* 2018, 731, 478–488.
157. Ciric′, A., Aleksic′, J., Barudžija, T., Antic′, Ž., Đordevic′, V., Medic′, M., Periša, J.B., Zekovic′, I., Mitric′, M., Dramicanin, M.D. Comparison of three ratiometric temperature readings from the Er^{3+} upconversion emission. *Nanomaterials (Basel)* 2020, 10, 4(627).
158. Rakov, N., Maciel, G.S. Exploring the $^4I_{13/2}\rightarrow{}^4I_{15/2}$ radiative transition from Er3+ in Y_2O_3 for temperature sensing. *Journal of Luminescence* 2018, 199, 293–297.
159. Miniajluk, N., Boulesteix, R., Deren, P.J. Spark plasma sintering of double perovskite Ba_2MgWO_6 doped with Ce^{3+}: Part I—Structural and microstructural characterizations. *Ceramics International* 2020, 46(6), 7602–7608.
160. Moyez, A. Sk., Roy, S. Dual-step thermal engineering technique: A new approach for fabrication of efficient $CH_3NH_3PbI_3$-based perovskite solar cell in open air condition. *Solar Energy Materials and Solar Cells* 2018, 185, 145–152.
161. Roy, S., Datta, S. Applications of polymers in perovskite solar cells: A review. *Annals of Chemical Science Research* 2020, 2(2), 1–4.
162. Mishra, R., Bera, S., Chatterjee, R., Banerjee, S., Bhattacharya, S., Biswas, A., Mallick, S., Roy, S. A review on Z/S—scheme heterojunction for photocatlytic applications based on metal halide perovskite materials. *Applied Surface Science Advances* 2022, 9, 100241.
163. Ding, Y., Maitra, S., Wang, C., Halder, S., Zheng, R., Barakat, T., Roy, S., Chen, L.-H., Su, B.-L. Vacancy defect engineering in semiconductors for solar light-driven environmental remediation and sustainable energy production. *Interdisciplinary Materials* 2022, 1(2), 213–255.
164. Moyez, A. SK., Roy, S. Thermal engineering of lead-free nanostructured $CH_3NH_3SnCl_3$ perovskite material for thin-film solar cell. *Journal of Nanoparticle Research* 2018, 20(5).
165. Moyez, A. SK., Mitra, S., Mukherjee, K., Sengupta, A., Roy, S. Structural features and optical properties of $CH_3NH_3Pb_{(1-x)}Sn_xCl_3$ Thin-film perovskites for photovoltaic applications. *Journal of Electronic Materials* 2020, 49, 7133–7143.
166. Datta, S., Roy, S. A review on perovskite nanostructure for renewable energy applictions in recent advances. *International Journal of Applied Nanotechnology* 2020, 6(2), 1–4.
167. Roy, S., G.Botte, G. Perovskite solar cell for photocatalytic water splitting with a TiO_2/Co-doped hematite electron transport bilayer. *RSC Advances* 2018, 8(10), 5388–5394.
168. Dey, A., Karan, P., Sengupta, A., Moyez A. SK., Sarkar, P., Majumdar, B.S., Pradhan, D., Roy, S. Enhanced charge carrier generation by dielectric nanomaterials for quantum dots solar cells based on CdS-TiO_2 photoanode. *Solar Energy* 2017, 158, 83–88.

169. Kajal, P., Verma, B., Vadaga, S.G.R., Powar, S. Costing analysis of scalable carbon-based perovskite modules using bottom up technique, *Global Challenges* 2021, 6, 2100070.
170. Čulík, P., Brooks, K., Momblona, C., Adams, M., Kinge, S., Maréchal, F., et al. Design and cost analysis of 100 MW perovskite solar panel manufacturing process in different locations. *ACS Energy Letters* 2022, 7, 3039–3044.
171. Su, R., Xu, Z., Wu, J., Luo, D., Hu, Q., Yang, W., et al. Dielectric screening in perovskite photovoltaics. *Nature Communications* 2021, 12, 2479.
172. Li, Z., Yang, M., Park, J.-S., Wei, S.-H., Berry, J.J., Zhu, K. Stabilizing perovskite structures by tuning tolerance factor: Formation of formamidinium and cesium lead iodide solid-state alloys. *Chemistry of Materials* 2016, 28, 284–292.
173. Serpetzoglou, E., Konidakis, I., Kakavelakis, G., Maksudov, T., Kymakis, E., Stratakis, E. Improved carrier transport in perovskite solar cells probed by femtosecond transient absorption spectroscopy. *ACS Applied Materials & Interfaces* 2017, 9, 43910–43919.
174. Bera, S., Saha, A., Mondal, S., Biwas, A., Malick, S., Chatterjee, R., Roy, S. Review of defect engineering in perovskites for photovoltaic application. *Materials Advances* 2022, 3(13), 5234–5247.

CHAPTER 2

Two-Dimensional Materials for Solar-Thermal Storage and Solar Power Conversion

AMIT KUMAR MISHRA[1] and SUNIL KUMAR[2*]

[1]*Department of Chemistry, Indian Institute of Technology, Kanpur, India*

[2]*Department of Chemistry, L.N.T. College, B.R.A. Bihar University, Muzaffarpur, Bihar, India*

*Corresponding author

ABSTRACT

The world's energy demand is increasing rapidly due to population growth and technological development. Solar energy is a promising and freely available green energy source for addressing long-term difficulties during an energy crisis. Low-cost functional materials for solar energy utilization are important and attractive. Two-dimensional (2D) nanomaterials have gained a great deal of attention owing to their unique optical and electrical features for next-generation solar devices. There are no known alternatives that possess a big specific surface area, excellent electrical conductivity, strong chemical stability, and flexibility. 2D materials' atomically thin bodies and high flexibility make them the ideal choice for synergistic integration with next-generation solar technologies. However, the structure and qualities of active materials have a significant impact on the overall performance of solar-thermal storage and conversion. 2D materials have abundant functional groups on the surface, which can greatly elevate solar affinity. Surface-active chromophores are highly susceptible to solar absorption. Due to their high carrier mobility, tunable band structure,

Advances in Energy Materials: Materials Fabrication, Properties and Applications.
Lishin T, Maciej Jaroszewski, and Sabu Thomas (Eds.)

DOI: 10.1201/9781779643674-2

and variable work function features, they considerably impact the charge carrier dynamics of solar systems. According to studies, the energy efficiency of solar conversion can be greatly improved by using 2D layered nanomaterials with phase change materials or other classical materials. This chapter outlines structural relevances and solar-thermal events in various emerging 2D materials for excellent thermal energy density and next-generation solar-based energy supply systems.

2.1 INTRODUCTION

Effective conversion, storage, and management of renewable energy sources and energy harvesting are the best ways to close the supply–demand gap and reduce environmental deterioration [1, 2]. The depletion of fossil fuels and climate change are two of the most pressing issues facing modern society [3, 4]. Renewable energy conversion and storage technologies offer a clean and sustainable method for reducing dependency on fossil fuels and associated environmental problems [5–7]. Nanostructured materials are distinct because their characteristics are entirely distinct from those of their bulk counterparts. Materials with nanoscale dimensions, such as polycrystalline materials with nanoscale crystallites, are nanostructured. Two-dimensional (2D) materials with novel structures and customizable properties have emerged as attractive possibilities for energy conversion and storage in recent years [5, 8]. Because of this, there is a lot of interest in the research and development of sustainable energy conversion and storage technologies [9, 10]. Even though there are many things that affect how well energy conversion and storage devices work, their overall performance depends heavily on the structure and properties of the materials they are made of. Several new nanomaterials (NMs) have been made to convert and store energy more efficiently. By changing the surface of nanoparticles, NMs' functionalization improves their properties and makes it possible for them to play a major role in the field of medicine. Also, the effect of surface functionalization on the nano-inclusions has a big effect on the increase in thermal conductivity and the long-term stability of the colloids [11, 12]. On the other hand, some applications need the surface of nanostructured materials to be functionalized. As researchers have found, surface functionalization is important for making graphene materials with good bulk and surface properties, which are needed for converting and storing energy efficiently [13, 14]. The world needs a lot more energy because its population is growing and its industries are changing [15]. It is important to remember that the world's population has grown by 2 billion in just one generation and that developing countries have done most of the work. One of the most common concerns of the 21st century is how to stop an energy crisis. For the future of the world, it is important to choose energy sources that are good for the environment. In this way, it is very important to think about renewable energy sources like solar energy, wind energy, hydropower, and geothermal energy, which are good for the environment. The industrial sector is no longer attracted to the usage of fossil fuels due to the rapid increase in conventional fuel prices and environmental restraints. By implementing renewable energy systems in industries, greenhouse gas emissions might be drastically decreased. Therefore, old energy sources should be replaced with renewable energy sources, and new technologies must be created and implemented in the industrial sector. However, solar energy may be the best alternative for the future due to a number of factors. First, solar energy is the most plentiful form of renewable energy, and the sun emits it at a rate of 3.8×10^{23} kW, of which the earth absorbs around 1.8×10^{14} kW [16]. The sun's energy reaches the planet in numerous forms, including light and heat. Compared with other temperate regions, Asia has the greatest capacity to receive solar radiation due to its high sunshine duration throughout the year. Second, the usage and tracking of solar energy have no negative effect on the ecosystem in which the natural balance is maintained for the benefit of living things. Third, due to its scalability and cost, a solar system can be used effectively for home, industrial, and community systems. A rush to develop solar energy is also underway due to the increasing dependency of the world's population on fossil fuels for a range of energy-related activities [17, 18]. Solar energy can be utilized through photothermal, photovoltaic (PV), and photocatalytic approaches [19, 20].

Since the discovery of graphene, the development of 2D materials is experiencing a renaissance in several fields. Graphene, transition metal dichalcogenides, transition metal carbides (TMCs), and other graphene-like elementary substances, such as black phosphorus (BP), silylene, etc., make up the majority of the rich family of 2D materials [21]. Comparing atomically thin 2D materials nanosheets to their bulk equivalents reveals several fascinating layering-dependent features due to quantum confinement and surface effects [22–24]. In this way, this chapter opens a new avenue to develop functional 2D nanostructures and composite characteristics as an energy storage medium to transform the efficiency of renewable energy technology.

2.2 RENEWABLE ENERGY AND ITS GLOBAL DISTRIBUTION

Conventional energy sources based on petroleum, coal, and natural gas have shown to be extremely successful economic growth generators. Due to the rapid depletion of conventional energy sources and rising energy demand, the global consumption of primary energy increased by 1.8% in 2012 [25]. As populations increase far more rapidly than the usual 2%, the demand for energy intensifies. In addition to global warming, problems with energy supply and consumption are associated with air pollution, acid precipitation, ozone depletion, forest destruction, and radioactive emission. If mankind wants to attain a prosperous energy future with minimal environmental repercussions, these challenges must be considered concurrently. The world's population is expected to double by the middle of the 21st century, and economic growth will almost certainly continue. Global demand for energy services is expected to rise by an order of magnitude by 2050, while primary energy demand will rise by 1.5–3 times. To address these issues and the impending energy shortage, much more use of renewable energy (Figure 2.1) sources and technologies is required [26]. Renewable energies are energy sources that are continually replenished by nature and derived directly from the sun (such as thermal, photo-chemical, and photo-electric), indirectly from the sun (such as wind, hydropower, and photosynthetic energy stored in biomass), or from other natural movements and mechanisms of the environment (such as geothermal and tidal energy). By 2050, the world may need 1000 EJ of primary energy if business-as-usual projections of global economic growth are to be realized. Given the problems facing both fossil and nuclear fuels, within a very few decades, renewable energy will have to account for most energy production [27].

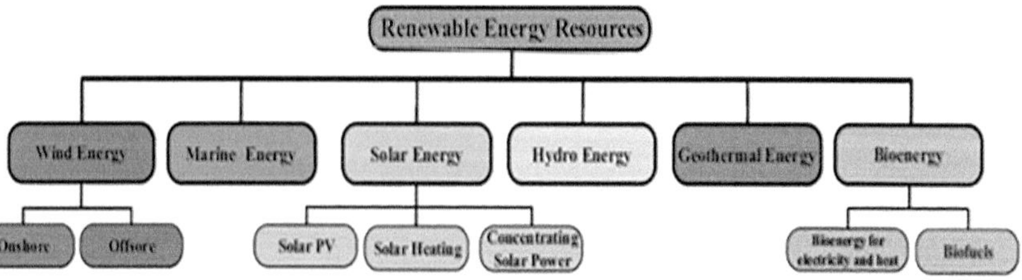

FIGURE 2.1 Classification of renewable energy sources.

2.3 SOLAR ENERGY

Solar energy is one of the most environmentally friendly energy sources that does not contribute to global warming. In one second, the sun emits more energy than humans have used since the beginning of time. Solar energy is frequently referred to as an "alternative energy" to fossil fuel energy sources like oil and coal. Solar energy is clearly more environmentally friendly than any other energy source, and it is the foundation of any serious sustainable development program. It does not deplete natural resources, emit CO_2 or other gaseous emissions into the atmosphere, or produce liquid or solid waste [28]. Solar energy is currently utilized in domestic and industrial settings for cooking, heating, cooling, lighting, space technology, and communications, among other applications. One of the biggest obstacles to the widespread adoption of solar energy is the high cost of PV module manufacture, as well as the presence of hazardous substances in some thin-film solar cells [29]. Due to the current global energy constraint and concomitant environmental problem, it is vital to find alternatives that are inexpensive, clean, and renewable. The conversion of solar energy using solar cell devices is regarded as one of the most exciting and promising forms of renewable energy. Recent research has focused on perovskite solar cells (PSCs), quantum dot (QD) solar cells, dye-sensitive solar cells, and other PV technologies [30–33]. As a result, the main challenge now is to figure out how to harvest solar energy more efficiently. PVs, which directly convert solar energy into electric power, is the most promising existing solar energy harvesting technology for dealing with the current energy crisis.

2.4 TWO-DIMENSIONAL NANOMATERIALS AS SOLAR-THERMAL FUELS AND SOLAR CONVERSION MEDIUM

2D structures, often known as nanosheets or thin films, are characterized by their atomic thickness, unique band structure, and vastly greater lateral dimensions. Due to quantum confinement, when layered materials are thinned to single and few layers, unusual (dramatically distinct from their parent 3D materials) and even novel electrical and optical properties appear during solar irradiation. Through the interlayer gap, 2D materials typically display strong in-plane covalent bonding and modest out-of-plane van der Waals interactions. Previously, it was mentioned that 2D materials can improve the PV performance or stability of PSCs [30]. Chemical vapor deposition, the electrochemical approach, and the oxidation–reduction method have been used successfully to open 2D materials. The guest species injected into the van der Waals gaps can be atoms, ions, or molecules. With an inherent bandgap and distinct crystal structures, these 2D materials have emerged as attractive candidates for a wide range of applications in next-generation electronics and optoelectronics science [8].

The solar-thermal conversion process of π-bonded carbon-based 2D having HOMO and LUMO orbitals permits electrons to move rapidly (Figure 2.2). In the π-conjugated structure, there are lots of π-bonded molecular orbitals and π-antibonded molecular orbitals in the molecular orbitals. After absorbing solar energy, the π electrons on the bonded molecular orbitals transition to the π antibonding molecular orbitals, at this time they are in an excited state. Then, the electrons release part of their energy in the form of heat and return back to the ground state, thereby producing a solar-thermal event.

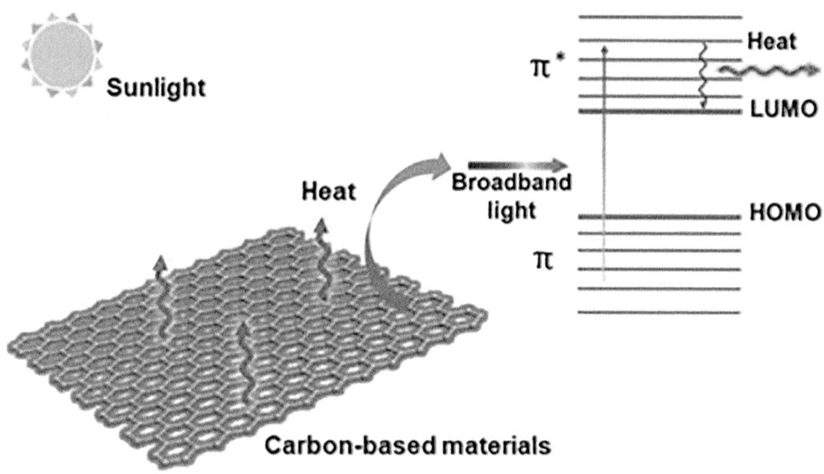

FIGURE 2.2 Solar-thermal conversion process in two-dimensional carbonaceous materials. *Source*: Reprinted with permission from Ref. [73].

2.4.1 *SOLAR-THERMAL CONVERSION*

Large amounts of latent heat can be reversibly absorbed and released during the phase transition process in phase-changing materials (PCMs), which are thermally responsive functional materials [34, 35]. As solar radiation intensity changes seasonally and geographically and therefore different PCM can be utilized to store solar thermal and the extent of phase change reflects the emission of energy. PCMs with a high latent heat and heat storage density are regarded as effective materials to address the timing mismatch between the heat supply and actual consumption. Solar-thermal energy storage efficiency can be assessed by following the empirical equation [36]:

$$\eta = \frac{m\Delta H}{APt}$$

where m indicates mass of the sample, ΔH is phase change enthalpy, A is the irradiated surface area, P is the power of solar radiation, and t stands for irradiation time.

However, due to their low solar absorbance and high reflectance, pristine PCMs are unable to drive solar-thermal conversion directly [35]. Photothermal materials serve as solar collectors, absorbing solar energy and transforming it into thermal energy (Scheme 2.1). In terms of cost and functionality, photothermal composite PCMs are ideal for storing renewable energy. A good approach in this regard is to incorporate photosensitizers (2D materials) into PCMs to enable effective broadband solar energy gathering. The following describes the basic solar-thermal conversion mechanism of composite PCMs. Electrons can travel from the ground state to the excited state using solar energy to excite photosensitive compounds. The excited electrons will then return to the ground state by quenching, energy transfer, or nonradiative relaxation [37]. The release of heat is an accompanying effect of these modes.

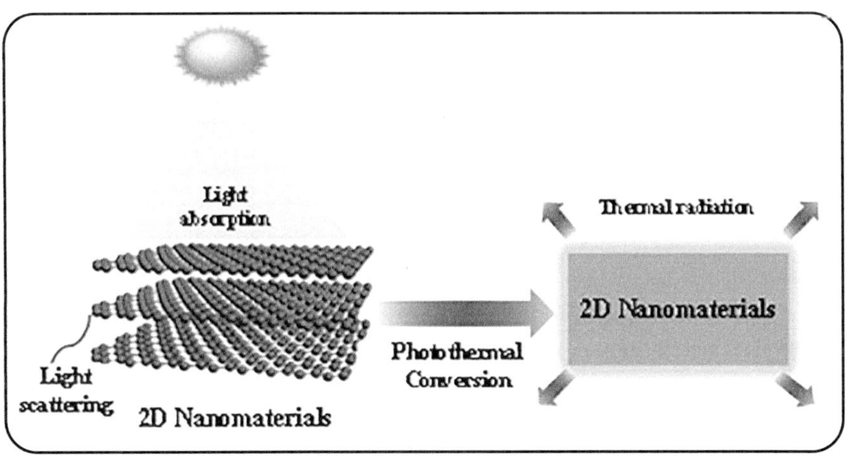

SCHEME 2.1 Schematic diagram of photothermal conversion into 2D nanomaterials (self-designed product).

2.4.2 PHOTOVOLTAIC CONVERSION

Using the PV effect, PV cells may convert solar energy into direct current electricity [38]. Solar PVs are high-performance energy conversion technologies in the modern age of science. Recent research has focused on 2D materials as a library of robust and effective cooperative materials in PSCs for high efficiency and stability. 2D materials serve as transparent electrodes, a perovskite layer additive, and a charge transport layer or its optimization in PSCs. The incorporated 2D materials improve PSC efficiency and stability in three ways: (1) by optimizing perovskite crystallization, (2) by promoting charge transport, and (3) by passivating defects [30]. Crystallinity, particle size, and morphology are three significant evaluation variables for perovskite layer quality. Excellent-quality perovskite layers (such as those with high crystallinity, preferential orientation, and increased grain size) are advantageous for achieving the greater performance of devices. Recent improvements in using 2D materials to change the substrate and control crystallization to improve the quality of the perovskite layer for PSCs that are most likely to be efficient and stable. The charge transport layer, which has been modified with graphene and its derivatives, has been chosen as the growing substrate for the preparation of high-quality perovskite layers. To increase the PV performance of devices, Chandrasekhar et al. [77] used nitrogen-doped graphene (NG)/ZnO composites as an electron transport layer and grew a substrate. To reduce charge recombination at grain boundaries, the perovskite films produced on NG/ZnO composites showed better crystallinity, high surface coverage, and increased grain size. In comparison to the reference ZnO-based devices (12.87%), the NG/ZnO-based PSCs thus achieved a higher PCE of 16.82%. Qin et al. [30] studied sulfated graphene oxide (sGO) combined with poly (3,4-ethylenediox- thiophene)–polystyrene sulfonate (PEDOT–PSS) and demonstrated how the grain size could reduce the perovskite film defect and increase the PCE from 11.5 to 13.9%, which is shown in Figure 2.3.

2.5 FUNCTIONAL 2D MATERIALS AND ITS STRUCTURAL CHARACTERISTICS FOR SOLAR ENERGY APPLICATION

2.5.1 GRAPHENE

Graphene, an allotrope of carbon in which the atoms are arranged in a honeycomb lattice, has gotten a lot of attention because of how interesting

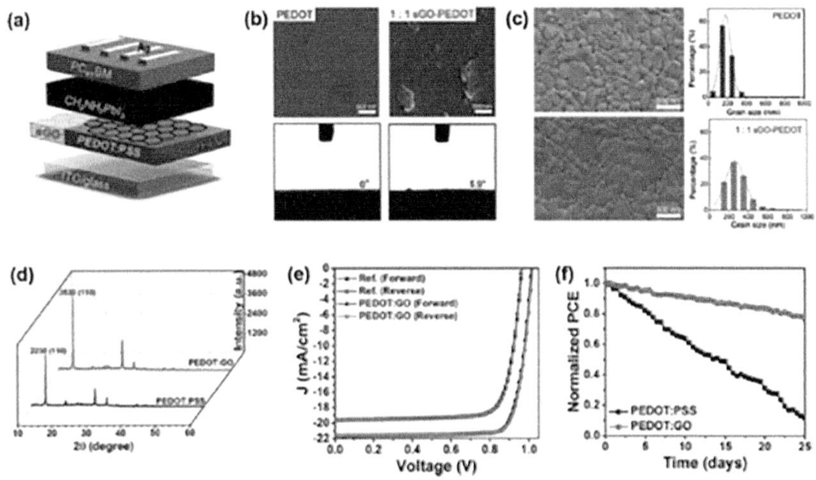

FIGURE 2.3 (a) Schematic of the device configuration of inverted PSCs with the sGO/PEDOT composite as the growing substrate. (b) Comparison of scanning electron microscopy (SEM) images and contact angles of the perovskite precursor solutions between PEDOT and sGO/PEDOT films. (c) SEM images and distributions of the perovskite grain size deposited on PEDOT and sGO/PEDOT substrates. (d) X-ray diffraction (XRD) patterns of perovskite prepared on the PEDOT: PSS substrate with and without GO doping. (e) J–V curves of the champion PSCs on the PEDOT: PSS substrate with and without GO doping. (f) Stability of the PSCs on the PEDOT: PSS substrate with and without GO doping under air conditions.

Source: Reprinted with permission from Ref. [30].

and unusual it is as a material. Since it was discovered, graphene has been the most studied 2D material because it has so many interesting properties, such as a high theoretical surface area (2600 $m^2\ g^{-1}$), excellent thermal conductivity (4840–5300 $Wm^{-1}\ K^{-1}$), extreme mechanical properties (tensile strength up to 130 GPa, elastic modulus of 1000 GPa), and ultra-high electron mobilities ($2 \times 10^5\ cm^2\ V^{-1}\ s^{-1}$). Graphene is made up of sp^2-hybridized carbon atoms that are stuck together in a plane that is only one atom thick and has a pattern of hexagons that repeat [39]. Graphene has been used extensively in energy-efficient applications to realize high-performance, efficient, low cost, lightweight, and flexible devices to meet the world's limited energy resources [8]. Graphene can be produced via a variety of bottom-up and top-down techniques, including repetitive peeling, sonication, ball milling, exfoliation, chemical vapor deposition, silicon evaporation, epitaxial growth, unzipping multiwall carbon nanotubes, and electric arcs [40]. Although there have been substantial breakthroughs, it is still very difficult to produce affordable, high-quality graphene on an industrial scale and over a wide area.

2.5.2 GRAPHENE OXIDE

Graphite oxide has a layered structure similar to graphite, but the plane of carbon atoms in graphite oxide is heavily decorated with oxygen-containing groups, which not only increases interlayer distance but also makes the atomic-thick layers hydrophilic [41]. As a result, these oxidized layers can be exfoliated using moderate ultrasonication in water. If the exfoliated sheets contain only one or a few layers of carbon atoms, they are referred to as graphene oxide (GO). GO is derived from a highly oxidized form of the graphene molecule with the use of numerous powerful oxidizing chemicals. Particularly, GO is chemically exfoliated from other graphene derivatives due to its superior surface functionality, amphiphilicity, aqueous appearance, fluorescence quenching ability, and surface-enhanced Raman scattering characteristics. These unique properties of GO are a result of the short sp^2 carbon domains that are surrounded by sp^3 carbon domains and the oxygen-containing hydrophilic functional groups in their unusual chemical structure. Large oxygenic surface functionalities enhance the solar energy storage efficiency. Moreover, degree of surface oxidation decides the capacity of GO toward solar energy storage.

2.5.3 REDUCED GRAPHENE OXIDE

Graphene, along with its derivatives GO and reduced GO (rGO), is one of the most well-known carbon compounds [42]. Instead, GO can be manufactured quickly and cheaply in the desired quantities [43]. GO has a 2D structure similar to graphene, but the single layer of carbon atoms is covalently functionalized with oxygen-containing groups (hydroxyl, epoxide, carbonyl, etc.) produced via chemical exfoliation of graphite flakes on the basal plane and on the edges. These functional groups provide GO with high colloidal stability in the water as well as a distinctive set of mechanical, colloidal, or optical properties, making it simple to process in dispersions. Fortunately, GO can recover graphene-like characteristics with subsequent reductive exfoliation processes, which convert GO to rGO. At present, rGO is an excellent compromise between graphene and GO. This is not only because rGO has graphene-like properties, including reasonably strong conductivity, but also because it is straightforward to manufacture in appropriate amounts from inexpensive GO utilizing a range of (electro) chemical, microwave-assisted thermal, and photo-assisted thermal techniques. It is critical to understand that the variety of reduction processes will result in different quality rGO with (significantly) altered properties, depending on the degree of reduction [44].

2.5.4 BLACK PHOSPHOROUS

BP, a rising star 2D material, has recently sparked a research boom. The publication of almost 400 BP-related studies since 2014 demonstrates the rational importance and fascination at the forefront of this research. Recent interest in phosphorene has been driven by its exceptional physical features and potential for interesting applications. Phosphorene, a mono-layer or multilayer BP that is similar to graphene, can be manufactured from bulk BP by a variety of processes. Phosphorus's most stable allotropy under ambient conditions is BP. The direct tunability of the band gap [45] from 0.3 eV (in bulk) to 1.5 eV (in monolayer), improved charge carrier mobility, and anisotropic in-plane characteristics are some of the many benefits of BP. These characteristics make BP an exceptional choice for nanophotonic device development using 2D layered materials and modest band gap-based nanoelectronics applications [46]. It has a large photon absorption window from the visible to the mid-infrared range of the solar spectrum, making it an excellent choice to serve as a photocatalyst [47]. Functional 2D BP nanostructures, such as surface-modified 2D BP and mixed-dimensional 2D BP-based heterostructure, have been carefully designed and successfully synthesized for next-generation devices with high device performance and excellent stability. This will allow its users to go beyond its current limits and help it reach its full potential [48].

2.5.5 HEXAGONAL BORON NITRIDE

Due to its intriguing properties and unique physical phenomena, hexagonal boron nitride (h-BN), also known as "white graphene," is one of the most popular 2D materials [49]. The geometry of h-BN's atomic structure is similar to that of graphene, with boron and nitrogen atoms replacing carbon atoms at the vertices of the planar hexagonal sp^2 network and van der Waals forces pulling adjacent h-BN layers together. With the rapid development of 2D materials, more novel properties and potential applications for h-BN have recently been discovered. Because of its low electron delocalization, h-BN is electrically insulating, with a large energy gap of about 5.96 eV [50]. Because of their layered structure and high chemical and thermal stability, they can be used as lubricants and protective coatings. Unlike graphene, which has no gaps and many useful electronic properties, h-BN is an insulator because of its wide band gap. This makes it hard to use in the fields of energy conversion and storage. Researchers have looked into how to change the properties of h-BN so that it can be used in many different ways. Both doping with different kinds of atoms and grafting functional groups have been shown to work [49]. In terms of optoelectronics and photonics, h-BN's natural hypertonicity and compelling photonic properties make it ideal for use in low-energy optoelectronic devices in the ultraviolet and infrared ranges [51]. Exfoliated 2D h-BN nanosheet's appealing features, such as strong thermal conductivity, chemical inertness to acid, and outstanding oxidation resistance up to 850°C, are also employed to build composite PCMs for thermal energy storage [52].

2.5.6 GRAPHITIC CARBON NITRIDE (g-C_3N_4)

At ambient conditions, graphitic carbon nitride (g-C_3N_4) is regarded as the most stable nitrogen-containing allotrope. Similar to graphite, g-C_3N_4 is a metal-free layered conjugated material in which van der Waals force holds the stacking layers (covalent C–N bonds) and each layer is composed of tri-s-triazine units connected with planar amino groups [53]. The tri-s-triazine ring structure provides high thermal stability (600°C in air) and chemical stability in both acidic and alkaline environments. The g-C_3N_4-based NMs have excellent electronic band structures, electron-rich properties, basic surface functionalities, high physicochemical stabilities, and are "earth-abundant" [54]. This is a good, visible-light-driven semiconductor used in various energy conversion devices.

2.5.7 MXENES

MXenes are a type of 2D inorganic compound composed of TMCs, nitrides, or carbonitrides with only a few atomic layers in thickness. Mn + 1XnTz (n = 1, 2, 3) is the general formula for MXenes, where M is the early transition metal (such as Sc, Ti, Zr, Hf, V, Nb, Ta, Cr, Mo, and so on), X is carbon and/or nitrogen, and T represents surface termination (e.g., OH, O, and F) [55]. MXene's interesting properties have led to a wide range of uses and a lot of interest in many fields. MXene-based materials were used a lot in catalysts, ion sieving, photothermal conversion, topological insulators for field-effect transistors, and hydrogen evolution reactions, where they worked very well [56]. This 2D layered MXene is similar to graphene. MXene has the same properties as rGO and GO. Also, MXene-based PCMs and nanofluids have shown significant potential for thermal energy storage [50].

2.5.8 BOROPHENE

As carbon's neighbor on the periodic table, boron (B) has many of the same properties as carbon (C). For example, they both have similar valence orbitals and can use sp^2 hybridization to make low-dimensional structures. Depending on how big they are, B clusters can have almost flat, flat, or cage-like shapes [57]. Researchers have been interested in the possibility of making a B monolayer or borophene for a long time showing that the chemistry of B and C clusters is similar in some ways. Borophene is naturally metallic, has high flexibility, properties that can be changed and can superconduct at high temperatures [58]. Borophene, a single sheet of boron, is regarded as an even more remarkable NMs than graphene. In contrast to graphene, which is an isotropic material, borophene exhibits considerable anisotropy due to the particular orientation of B atoms inside its lattice structure. In addition, the electrical and transport properties of borophene are largely determined by the edge arrangement direction. In 2D-borophene, strong phonon–phonon scattering results in a low thermal conductivity. In addition, because of its particular capabilities and outstanding carrier mobility, borophene is an excellent electrode material for Na–ion batteries, Li-ion batteries, and Mg-ion batteries. Borophene has a metallic character that is highly anisotropic due to its peculiar band topologies. Borophene can be utilized in rechargeable batteries as an anodic material due to its substantial ion storing capacity. Numerous uses exist for these devices in sensor development, energy storage, and other fields. Borophene can exist in a structural configuration similar to graphene, yet β_{12}, χ_3, and β_6 polymorphs of borophene can be experimentally prepared [59].

2.5.9 METAL-ORGANIC FRAMEWORK

Metal-organic frameworks (MOFs), commonly referred to as porous coordination polymers, are a kind of crystal composed of metal ions/clusters and organic ligand linkers by means of coordination bonds. The benefits of MOFs include a high surface area, well-defined structures, a variety of chemical compositions, customizable pore topologies, varied production, and numerous functionalities [60]. MOF materials are new porous materials with a large surface area, pore size and volume, and variable pore geometry, resulting in a greater adsorption capacity than silica gel and zeolite. MOFs are innovative materials with a high promise for energy storage and cooling applications via sorption [61]. MOFs, which are hybrid organic/inorganic crystalline solids, are extremely porous materials with a wide range of compositions and structures. Energy conversion systems based on photocatalysis and electrocatalysis are among the most environmentally friendly and renewable methods of addressing energy and environmental issues. Photocatalysis, which involves light harvesting, electron–hole separation, and surface redox processes, can produce a variety of high-reward molecules such as H_2 and O_2 through water splitting and carbonaceous fuels through CO_2 reduction. Many ECS systems rely on electrocatalysis, such as water splitting, electrocatalytic reduction of N_2 and CO_2, fuel cells, and metal-air batteries. These energy storage devices enable efficient reversible electricity storage and release for portable electronics, electric vehicles, and power grids. However, many challenges remain in these fields. Of particular interest, MOFs have emerged as promising platforms for developing advanced materials for efficient ECS systems (MCOF). In particular, because of their unique structural and electrical properties, 2D covalent organic frameworks (COFs), a recently discovered class of 2D NMs, have drawn special attention. MOFs have several unique compositional and structural advantages over conventional materials due to the highly ordered and tunable metal nodes and organic linkers. MOFs' excellent gas storage capacity at room temperature has fueled their rapid growth, resulting in a plethora of new adsorbents with potential applications in vehicle gas tanks, fuel cells, and stationary power facilities [62]. Nevertheless, pure MOFs often have poor mechanical strength, chemical/electrochemical stability, and electric conductivity, which significantly limit their direct application in ECS systems. When MOFs and one or more functional materials are combined, the underlying topology of the MOFs is preserved, and all the components are used to their full potential with synergistic effects for improved ECS performances [60].

2.5.10 COVALENT–ORGANIC FRAMEWORKS

The COFs are emerging 2D nanostructures that possess appropriate frameworks with large and abundant pore surfaces, which contribute to their good thermochemical stability, large surface areas, and abundant active functional building moiety. Based on its structural versatility and tunability, COFs could provide a platform for developing functional materials to saturate multi demands, such as photoelectronics and energy implications. Lv et al. [63] developed naphthalenediimide and triphenyl-amine moieties (NT-COF) with 2D porous nanosheets and high specific surface area of 1276 $m^2 g^{-1}$ to utilize as cathode active materials in solar Li ion battery. These types of 2D materials open a door to convert solar energy into electrochemical energy via reversible charge transport across available redox-active centers.

2.5.11 TRANSITION METAL DICHALCOGENIDES

Over the past few decades, researchers have become more and more interested in transition metal chalcogenides (TMCs, mostly sulfides or selenides) as potential electrode materials for energy storage and conversion. This is because TMCs have tunable stoichiometric compositions, unique crystal structures, and many redox sites, as well as higher electrical conductivity than transition metal oxides [64]. Each transition metal molecule is covalently bound to six chalcogenides (such as tellurides, sulfides, and selenides), and the separate layers are held together by van der Waals force. Size and thickness are freely modifiable, allowing for more synthetic techniques to combine optical, electrical, and electronic phenomena [65].

2.6 CHROMOPHORE GROUPS, SURFACE DEFECT AND QUANTUM CONFINEMENT OF 2D NANOMATERIALS

The active chromophore (functional group) emerges the photostorage capacity of the 2D nanostructures. Figure 2.4 demonstrated various 2D layered structures that offer suitability and selectivity to regulate the solar science. The energy structures are highly prone to extend the absorption window of solar radiation. The term "quantum confinement" mostly refers to the energy of electrons that are contained (electrons or electron holes). Compared to bulk materials, the energy levels of electrons in nanocrystals will not be continuous. Such effects occur when the dimensions of the potential approach the de Broglie wavelength of electrons, causing shifts or distinct energy levels. The phenomena are termed quantum confinement, and nanocrystals are frequently referred to as QDs [57]. In addition, this QD impact influences NM features such as the material's electrical, optical, and mechanical behavior. NMs have electrons with greater energy levels than bulk materials due to their unique properties.

Depending on the size of the QD, confined electrons have more energy than bulk electrons. When their dimensions are reduced from 2D to 1D or 1D to 0D, semiconductor nanoparticles exhibit intriguing features.

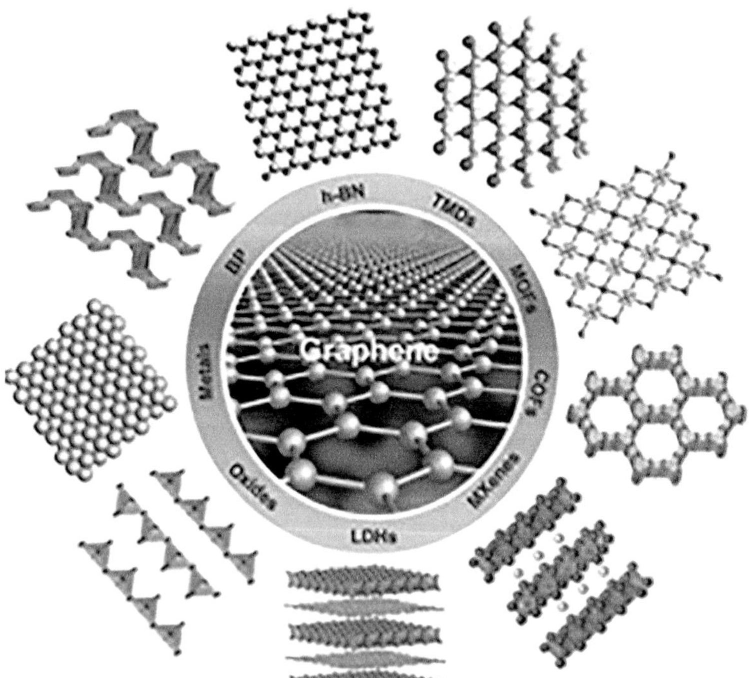

FIGURE 2.4 Schematic illustration various kinds of 2D layered materials.

Source: Reprinted with permission from Ref. [74].

The quantum confinement effect may occur when the size and shape of NMs are reduced to 100–10 nm or less. Understanding the phenomenon of QDs is crucial to learn more about quantum confinement (QDs). Quantum confinement effects are visible in a new family of materials called QDs [66]. Both molecules and QDs, which are extremely small semiconductor crystals on the order of nanometers in size, are tightly contained electrons or electron–hole pairs known as "excitons." As previously indicated, group II–VI, III–V, and IV–VI-based materials are the only ones that provide QD with its unusual tunable band gap features. With respect to variations in particle size, the optical, electrical, and bandgap properties are tunable, resulting in a variety of applications.

2.6.1 ADVANCEMENT OF 2D NANOMATERIALS, ENERGY DEVICE AND THEIR CONVERSION EFFICIENCY

In addition, several inherent difficulties of 2D NMs remain formidable obstacles when confronted with specific application needs. For instance, the severe layer restacking and aggregation that occurs during electrode fabrication and electrochemical response hinder the performance of 2D NMs. Furthermore, the surface of some 2D NMs is chemically inert and lacks a strong affinity for target ions. As a result, their electrochemical performance would be inadequate. To overcome this limitation, various functionalization strategies, such as heteroatom doping, defect engineering, pore construction, and so on, are proposed to build a series of novel 2D NMs. Because of the efficacy of these strategies, electrodes composed of 2D NMs exhibit excellent electrochemical performance, promoting the development of these advanced energy storage systems [67]. TEM images as shown in Figure 2.5a also directly confirm the 2D sheet-like morphology of WOx nanosheets. The morphological size sample did not change after introducing the oxygen vacancies or defects. The solar-thermal generation efficiency of 2D defective WOx nanosheets mainly relies on their oxygen defects. The solar evaporation rate and evaporation efficiency of resultant sheet under certain solar illumination intensity are shown in Figure 2.5b. The elevation of irradiation intensity can increase the solar-thermal generation efficiency of oxygen vacant nanosheet; however, there is a maximum and stable solar-thermal conversion efficiency that has been assessed for this system. The solar absorption capacity can be tuned by the concentration of oxygen defects. The solar evaporation efficiency of

2D defective WOx nanofluids is as high as 78.6% under 1 sun irradiation [68]. The uniform stacked Cu_3BiS_{3s}–Ti_3C_2 composite material has a steady-state temperature of approx. 62.3°C under one sun incident light intensity and highly prone for near-infrared absorption (Figure 2.5c and d). Thickness-dependent solar characteristics confirm the continuous elevation of steady-state temperature proving that resultant 2D anchored materials achieve higher surface temperature at 0.48 and 0.60 mg·cm^{-2} thickness. It guarantees a reasonably high steam generation rate (1.32 kg m^{-2} h^{-1}) even with a thinner dimensions (0.48 mg cm^{-2}) and a comprehensive conversion efficiency higher than 90% [69].

2D WTe_2/PtXY (X, Y = S, Se) donor–acceptor heterostructure have strong solar absorption coefficients (1 × 10^5 cm^{-1}) in the visible light region (Figure 2.6a). Heterostructure establishes suitable band structure and interlayer interaction to induce new optical transitions. The PCEs of 2D WTe_2/PtXY systems, that is, WTe_2/PtS_2, WTe_2/$PtSe_2$, WTe_2/SePtS, and WTe_2/SPtSe heterostructure are computed as 11.24, 21.43, 19.86,

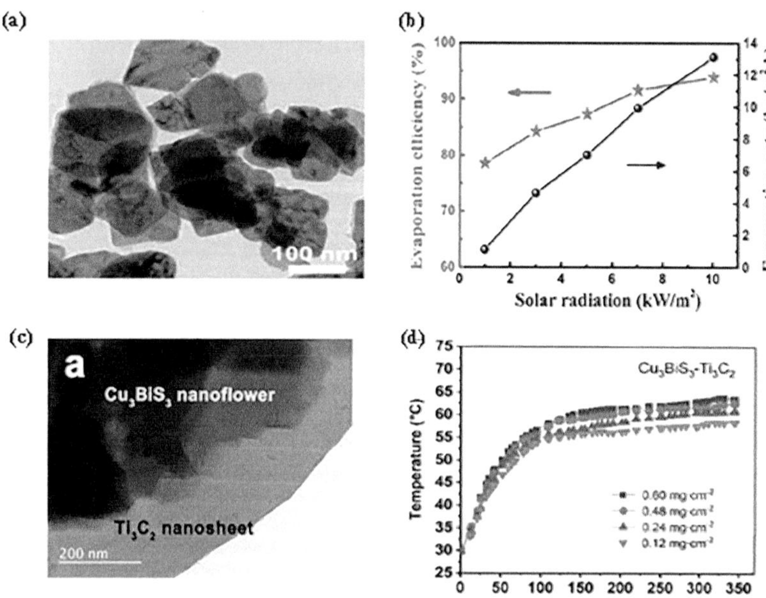

FIGURE 2.5 (a) TEM image of oxygen vacant WOx 2D nanosheet. (b) Solar-thermal behavior of resultant 2D nanomaterials [68]. (c) TEM image of 2D (Mxene) supported Cu_3BiS_3-Ti_3C_2 materials (d) steady-state temperature curve of samples with different thickness. *Source*: Reprinted with permission from Ref. [69].

and 10.29%, respectively, as shown in Figure 2.6b. The donor WTe_2 has a band gap of 1.34 eV, which is suitable for the absorption of the solar spectrum. Meanwhile, the conduction band offset of WTe_2/PtSe2 system develops the smallest conduction band offset (0.23 eV) among other systems [70]. The functionalization of 2D NMs provides viable ways to overcome their inherent limitations and, more crucially, allows for the tailoring of the target features for individual applications. The absorption coefficient can be improved by synergistic integration into 2D–2D systems (Figure 2.6c and d). This chapter presents an overview of recent scientific advancements in the functionalization of chemically produced graphene for enhancing the capacitance of ECs. Various methods of covalent and noncovalent functionalization have been devised for creating graphene materials with the desired structures and characteristics. Unfortunately, the lack of a bandgap, ease of aggregation, and

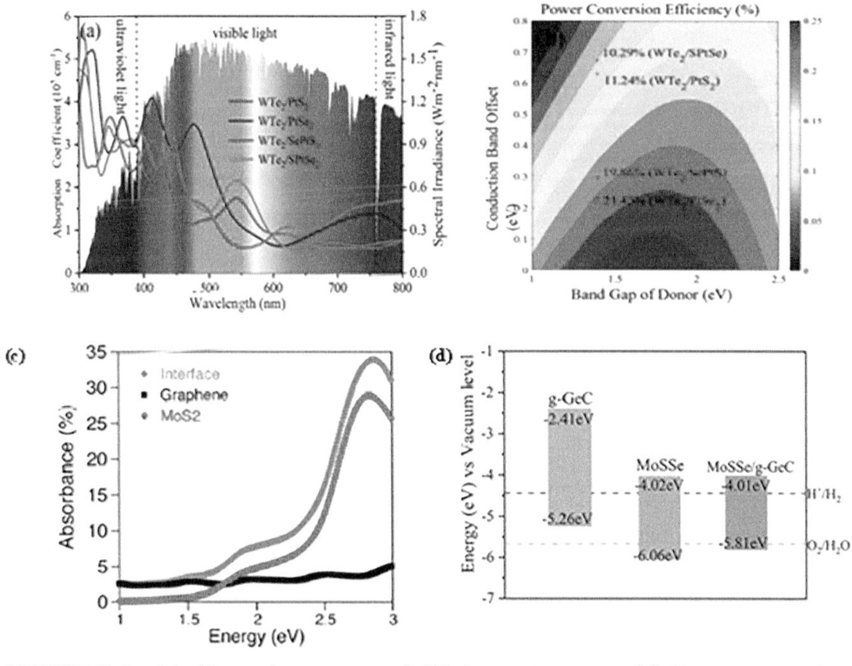

FIGURE 2.6 (a) Absorption spectra of 2D heterostructures. (b) Power conversion efficiency for solar systems (devices) having family of heterostructures [70]. (c) Absorption spectra of MoS_2/graphene interface and its compromising monolayers [75]. (d) Band alignment into 2D materials and its heterostructure.

Source: Reprinted with permission from Ref. [76]. Copyright © 2014 Elsevier Ltd.

poor dispersion in common solvents make it difficult to directly apply 2D pristine graphene sheets. As a result, there is ongoing research interest in transforming graphene sheets into functionalized graphene structures of various dimensions [71]. The functionalization of graphene is produced chemically to increase ECs' capacitance. Analysis of the role that graphene can play in improving/enhancing the electrochemical properties of pseudocapacitive materials (e.g., transition metal oxides/hydroxides and conducting polymers) for developing pseudocapacitors focuses on strategies/methods for minimizing the restacking of reduced GO to maximize the specific surface area available for accommodating

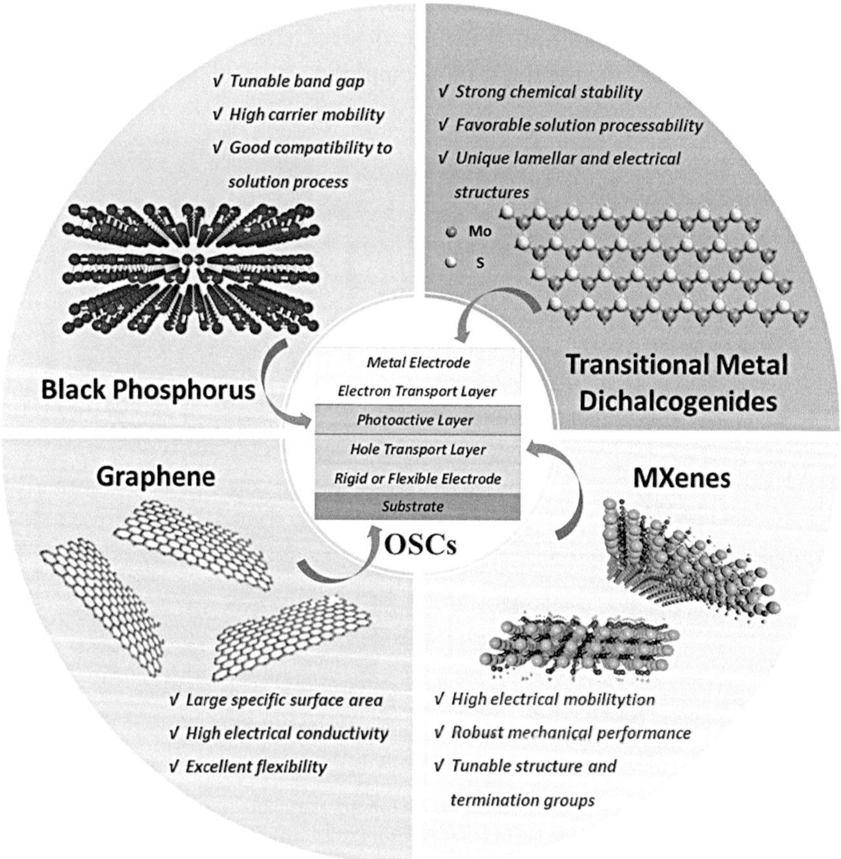

FIGURE 2.7 2D nanomaterials and their specialties in organic solar cells applications.
Source: Reprinted with permission from Ref. [33].

charge (energy) [72]. Various 2D nanostructures have been shown in Figure 2.7 which inspires to develop energy-active functional materials and advanced energy devices.

Based on variety of applications in solar PVs and energy storage devices, energy active 2D nanostructures are highly promising innovations in the field of energy science and technology.

2.7 CONCLUSION

Solar energy is one of the renewables with the most potential. Solar energy can be used in many ways, such as through solar thermal and solar PV technologies. Overall efficiency of the system depends on integration and interfacial charge transport in solar collectors (absorbers). Since graphene was founded, there has been a growing interest in 2D materials demonstrating their efficacy as solar-thermal storage and conversion devices. Incorporating nanotechnology into the films has particular promise for increasing solar energy conservation efficiency and decreasing production costs. High-performance 2D NMs possess unique solar radiation absorbance, extended thermal conductivity, and a localized surface plasmon phenomenon, leading to broadband (ultraviolet-visible-infrared) light absorbance and excellent photothermal transformation. This chapter unravels the importance of various advanced 2D materials/NMs for boosting the efficiency of solar-thermal storage and conversion and thus, outfits the futuristic solar technologies for the practical energy demand of next generations.

2.8 FUTURE PERSPECTIVES

2D layered NMs have been in the forefront of solar cell industry due to their unique opto-electric properties. Nevertheless, 2D materials integrated solar system performance might be degraded by environmental factors, that is, reactive gases and humidity, signifying the need for protective layers. In the last decade, scientists have used 2D h-BN in various solar devices to passivate the surface of active materials or design the interface of heterojunction solar conversion systems. The peculiar characteristics, large bandgap, excellent chemico-thermal stability, excellent crystallinity, medium dielectric constant, and easy transfer to substrates without affecting its bulk properties make 2D potential candidates. Even though the physical and chemical properties of the widest range of 2D materials are very different, the uses of each 2D material are still limited by its single property. By making 2D materials more accessible, they show that they have a huge amount of potential for a wide range of fundamental and technological applications. Graphene materials, however, need to have their surfaces functionalized to meet the needs of certain applications (like converting and storing energy). With the progress of physical and chemical changes, functionalized materials have unique properties like tuneable band gaps and high porosity that allow them to be used in different energy-related ways. Recent studies have demonstrated that 2D layered materials significantly affect the efficiency of organic solar cells by strengthening the operational stability of the equipment. The remarkable solar-thermal conversion efficiency of composite PCMs containing 2D NMs may open the advance approach of energy science and technology for sustainable solar utilization at low power consumption.

ACKNOWLEDGMENT

Dr. Amit Kumar Mishra acknowledges the technological support of Department of Chemistry, IIT Kanpur during chapter design and management. Dr. Sunil Kumar is thankful to L.N.T. College (B.R.A. Bihar University), Muzaffarpur for supporting to publish this work at research ground.

CONFLICTS OF INTEREST

The authors have declared no conflict of interest.

KEYWORDS

- **2D materials**
- **surface active chromophore**
- **solar-thermal energy**
- **solar energy storage**
- **solar technology**
- **energy efficiency**

REFERENCES

1. López-Sabirón, A.M., Royo, P., Ferreira, V.J., Aranda-Usón, A., Ferreira, G. Carbon footprint of a thermal energy storage system using phase change materials for industrial energy recovery to reduce the fossil fuel consumption. *Applied Energy* 2014, 135, 616–624. doi:10.1016/j.apenergy.2014.08.038.
2. Kumar, S., Maurya, I.C., Prakash, O., Srivastava, P., Das, S., Maiti, P. Functionalized thermoplastic polyurethane as hole conductor for quantum dot-sensitized solar cell. *ACS Applied Energy Materials* 2018, 1, 4641–4650. doi:10.1021/acsaem.8b00783.
3. Yadav, P.K., Kumar, S., Maiti, P. Conjugated polymers for solar cell applications, in: V. Kumar, K. Sharma, R. Sehgal, S. Kalia (Eds.), *Conjugated Polymers for Next-Generation Applications, Volume. 2 Energy Storage Devices*, Woodhead Publisher (An imprint of Elsevier), 2022, pp. 367–401. doi:10.1016/B978-0-12-824094-6.00004-2.
4. Prakash, R., Kumar, S., Maiti, P. Carbon nanotube based nanomaterials for solar energy storage devices, in: G. Manik, S.K. Sahoo (Eds.), *Current and Future developments in Nanomaterials and Carbon Nanotubes: Applications of Nanomterials in Energy Storage and Electronics*, 1st edn., Bentham Science, 2022, pp. 1–18. doi:10.2174/97 898150507141220300004.
5. Tao, H., Fan, Q., Ma, T., Liu, S., Gysling, H., Texter, J., Guo, F., Sun, Z. Two-dimensional materials for energy conversion and storage. *Progress in Materials Science* 2020, 111, 100637. doi:10.1016/j.pmatsci.2020.100637.
6. Saraswat, A., Kumar, S. A topical study of electrochemical response of functionalized conducting polyaniline: An overview. *European Polymer Journal* 2023, 182, 111714. doi:10.1016/j.eurpolymj.2022.111714.
7. Kumar, S., Prakash, R., Maiti, P. Advanced batteries and charge storage devices based on nanowires, in: G. Manik, S.K. Sahoo (Eds.), *Curr. Current and Future developments in Nanomaterials and Carbon Nanotubes: Applications of Nanomterials in Energy Storage and Electronics*, 1st edn., Bentham Science, 2022, pp. 159–175. doi:10.2174 /9789815050714122030012.
8. Das, S., Pandey, D., Thomas, J., Roy, T. The role of graphene and other 2D materials in solar photovoltaics. *Advanced Materials* 2019, 31, 1–35. doi:10.1002/ adma.201802722.
9. Kumar, S., Kumar, S., Rai, R.N., Lee, Y. Hong Chuong Nguyen, T., Young Kim, S., Van Le, Q., Singh, L. Recent development in two-dimensional material-based advanced photoanodes for high-performance dye-sensitized solar cells. *Solar Energy* 2023, 249, 606–623. doi:10.1016/j.solener.2022.12.013.
10. Kumar, S., Maiti, P. Review on functional electrolyte, redox polymers, and solar conversions in 3G emerging photovoltaic technologies: Progress and outlook. *Energy and Fuels* 2023, 37, 14473–14511. doi:10.1021/acs.energyfuels.3c01695.
11. Angayarkanni, S.A., Philip, J. Review on thermal properties of nanofluids: Recent developments. *Advances in Colloid and Interface Science* 2015, 225, 146–176. doi:10.1016/j.cis.2015.08.014.
12. Kumar, S., Prakash, R., Maiti, P. Redox mediation through integrating chain extenders in active ionomer polyurethane hard segments in CdS quantum dot sensitized solar cell. *Solar Energy* 2022, 231, 985–1001. doi:10.1016/j.solener.2021.12.043.

13. Kumar, S., Yadav, P.K., Maiti, P. Renewable cathode materials dependent on conjugated polymer composite systems, in: V. Kumar, K. Sharma, R. Sehgal, S. Kalia (Eds.), *Conjugated Polymers for Next-Generation ApplicationsVolume. 2 Energy Storage Devices*, 1st edn., Woodhead Publisher (An imprint of Elsevier), 2022, pp. 55–90. doi:10.1016/B978-0-12-824094-6.00002-9.
14. Kumar, S., Yadav, P.K., Prakash, R., Santra, A., Maiti, P. Multifunctional graphene oxide implanted polyurethane ionomer gel electrolyte for quantum dots sensitized solar cell. *Journal of Alloys and Compounds* 2022, 922, 166121. doi:10.1016/j.jallcom.2022.166121.
15. Kumar, S., Chaurasia, P.K., Patel, S.K.S. Improvement in the efficiency of organic semiconductors via molecular doping for OLEDs applications, in: L. Singh, R. Dubey, R.N. Rai (Eds.), *Organic Light Emitting Diode (OLED) Toward Smart Lighting and Displays Technologies*, 1st edn., CRC press, 2023, pp. 203–220. doi:10.1201/9781003260417-10.
16. Panwar, N.L., Kaushik, S.C., Kothari, S. Role of renewable energy sources in environmental protection: A review. *Renewable and Sustainable Energy Reviews* 2011, 15, 1513–1524. doi:10.1016/j.rser.2010.11.037.
17. Kannan, N., Vakeesan, D. Solar energy for future world: - A review. *Renewable and Sustainable Energy Reviews* 2016, 62, 1092–1105. doi:10.1016/j.rser.2016.05.022.
18. Kumar, S., Rai, R.N., D. Singh, Ansari, A.A., Lee, Y., Singh, L. Future perspectives on QDs embedded nano-fibrous materials as high capacity sustainable anode for Na-ion batteries technology. *MRS Energy Sustainability* 2023, 10, 238–246. doi:10.1557/s43581-023-00067-x.
19. Gong, J., Li, C., Wasielewski, M.R. Advances in solar energy conversion. *Chemical Society Reviews* 2019, 48, 1862–1864. doi:10.1039/c9cs90020a.
20. Saraswat, A., Kumar, S. Cutting-edge applications of polyaniline composites towards futuristic energy supply devices. *European Polymer Journal* 2023, 200, 112501. doi:10.1016/j.eurpolymj.2023.112501.
21. Tareen, A.K., Khan, K., Aslam, M., Liu, X., Zhang, H. Confinement in two-dimensional materials: Major advances and challenges in the emerging renewable energy conversion and other applications. *Progress in Solid State Chemistry* 2021, 61, 100294. doi:10.1016/j.progsolidstchem.2020.100294.
22. Xue, Y., Zhang, Q., Wang, W., Cao, H., Yang, Q., Fu, L. Opening two-dimensional materials for energy conversion and storage: A concept. *Advanced Energy Materials* 2017, 7, 1–23. doi:10.1002/aenm.201602684.
23. Kumar, S., Nanda Srivastva, A. Application of carbon nanomaterials decorated electrochemical sensor for analysis of environmental pollutants, in: *Analytical Chemistry—Advancement, Perspectives and Applications*, 1st edn., IntechOpen, 2021. doi:10.5772/intechopen.96538.
24. Kumar, S., Chaurasia, P.K. Functionalized nanomaterials based efficient photocatalyst for renewable energy and sustainable environment, in: A.S. et al Miridula Tripathi (Ed.), *Versatile Solicitations of Materials Science in Diverse Science Fields*, NOVA Science Publisher: New York, 2021, pp. 173–192.
25. Ellabban, O., Abu-Rub, H., Blaabjerg, F. Renewable energy resources: Current status, future prospects and their enabling technology. *Renewable and Sustainable Energy Reviews* 2014, 39, 748–764. doi:10.1016/j.rser.2014.07.113.

26. Dincer, I. Renewable energy and sustainable development: A crucial review. *Renewable and Sustainable Energy Reviews* 2000, 4, 157–175. doi:10.1016/S1364-0321(99)00011-8.
27. Moriarty, P., Honnery, D. What is the global potential for renewable energy?, *Renewable and Sustainable Energy Reviews* 2012, 16, 244–252. doi:10.1016/j.rser.2011.07.151.
28. Solangi, K.H., Islam, M.R., R. Saidur, Rahim, N.A., Fayaz, H. A review on global solar energy policy. *Renewable and Sustainable Energy Reviews* 2011, 15, 2149–2163. doi:10.1016/j.rser.2011.01.007.
29. Sagadevan, S. Recent trends on nanostructures based solar energy applications: A review. *Reviews on Advanced Materials Science* 2013, 34, 44–61.
30. Qin, Z., Chen, Y., Zhu, K., Zhao, Y. Two-dimensional materials for perovskite solar cells with enhanced efficiency and stability. *ACS Materials Letters* 2021, 3, 1402–1416. doi:10.1021/acsmaterialslett.1c00327.
31. Lu, H., Huang, Z., Martinez, M.S., Johnson, J.C., Luther, J.M., Beard, M.C. Transforming energy using quantum dots. *Energy & Environmental Science* 2020, 13, 1347–1376. doi:10.1039/c9ee03930a.
32. Nazeeruddin, M.K., E. Baranoff, Grätzel, M. Dye-sensitized solar cells: A brief overview. *Solar Energy* 2011, 85, 1172–1178. doi:10.1016/j.solener.2011.01.018.
33. Zhao, Y., Yu, L., Sun, M. Recent progress in emerging 2D layered materials for organic solar cells. *Solar Energy* 2021, 218, 621–638. doi:10.1016/j.solener.2021.02.066.
34. Mishra, A.K., Lahiri, B.B., Philip, J. Effect of surface functionalization and physical properties of nanoinclusions on thermal conductivity enhancement in an organic phase change material. *ACS Omega* 2018, 3, 9487–9504.
35. Mishra, A.K., Lahiri, B.B., Philip, J. Carbon black nano particle loaded lauric acid-based form-stable phase change material with enhanced thermal conductivity and photo-thermal conversion for thermal energy storage. *Energy* 2020, 191, 116572. doi:10.1016/j.energy.2019.116572.
36. Aftab, W., Khurram, M., Jinming, S., Tabassum, H., Liang, Z., Usman, A., Guo, W., Huang, X., Wu, W., Yao, R., Yan, Q., Zou, R. Highly efficient solar-thermal storage coating based on phosphorene encapsulated phase change materials. *Energy Storage Materials* 2020, 32, 199–207. doi:10.1016/j.ensm.2020.07.032.
37. Chen, X., Yu, H., Gao, Y., Wang, L., Wang, G. The marriage of two-dimensional materials and phase change materials for energy storage, conversion and applications. *EnergyChem* 2022, 4, 100071. doi:10.1016/j.enchem.2022.100071.
38. Wang, B., Iocozzia, J., Zhang, M., Ye, M., Yan, S., Jin, H., Wang, S., Zou, Z., Lin, Z. The charge carrier dynamics, efficiency and stability of two-dimensional material-based perovskite solar cells. *Chemical Society Reviews* 2019, 48, 4854–4891. doi:10.1039/c9cs00254e.
39. Kumar, R., Joanni, E., Singh, R.K., Singh, D.P., Moshkalev, S.A. Recent advances in the synthesis and modification of carbon-based 2D materials for application in energy conversion and storage. *Progress in Energy and Combustion Science* 2018, 67, 115–157. doi:10.1016/j.pecs.2018.03.001.
40. Olabi, A.G., Abdelkareem, M.A., Wilberforce, T., Sayed, E.T. Application of graphene in energy storage device—A review. *Renewable and Sustainable Energy Reviews* 2021, 135, 110026. doi:10.1016/j.rser.2020.110026.

41. Pei, S., Cheng, H.M. The reduction of graphene oxide. *Carbon* 2012, 50, 3210–3228. doi:10.1016/j.carbon.2011.11.010.
42. Ray, S.C. Application and uses of graphene oxide and reduced graphene oxide. *Applications Graphene Graphene-Oxide Based Nanomater* 2015, 39–55. doi:10.1016/b978-0-323-37521-4.00002-9.
43. Iliut, M., Gabudean, A.M., C. Leordean, Simon, T., Teodorescu, C.M., Astilean, S. Riboflavin enhanced fluorescence of highly reduced graphene oxide. *Chemical Physics Letters* 2013, 586, 127–131. doi:10.1016/j.cplett.2013.09.032.
44. Tarcan, R., Todor, O.-Boer, Petrovai, I., Leordean, C., Astilean, S., Botiz, I. Reduced graphene oxide today, *Journal of Materials Chemistry C* 2020, 8, 1198–1224. doi:10.1039/c9tc04916a.
45. Kiraly, B., Hauptmann, N., Rudenko, A.N., Katsnelson, M.I., Khajetoorians, A.A., Probing single vacancies in black phosphorus at the atomic level. *Nano Letters* 2017, 17, 3607–3612. doi:10.1021/acs.nanolett.7b00766.
46. Ren, X., Lian, P., Xie, D., Yang, Y., Mei, Y., Huang, X., Wang, Z., Yin, X. Properties, preparation and application of black phosphorus/phosphorene for energy storage: A review. *Journal of Materials Science* 2017, 52, 10364–10386. doi:10.1007/s10853-017-1194-3.
47. Liu, H., Hu, K., Yan, D., Chen, R., Zou, Y., Liu, H., Wang, S. Recent advances on black phosphorus for energy storage, catalysis, and sensor applications. *Advanced Materials* 2018, 30, 1–22. doi:10.1002/adma.201800295.
48. Wang, M., Zhu, J., Zi, Y., Wu, Z.G., Hu, H., Xie, Z., Zhang, Y., Hu, L., Huang, W. Functional two-dimensional black phosphorus nanostructures towards next-generation devices. *Journal of Materials Chemistry A* 2021, 9, 12433–12473. doi:10.1039/d1ta02027g.
49. Han, R., Liu, F., Wang, X., Huang, M., Li, W., Yamauchi, Y., Sun, X., Huang, Z. Functionalised hexagonal boron nitride for energy conversion and storage. *Journal of Materials Chemistry A* 2020, 8, 14384–14399. doi:10.1039/d0ta05008c.
50. Watanabe, K., Taniguchi, T., Kanda, H. Direct-bandgap properties and evidence for ultraviolet lasing of hexagonal boron nitride single crystal. *Nature Materials* 2004, 3, 404–409.
51. Gong, Y., Xu, Z.Q., Li, D., Zhang, J., Aharonovich, I., Zhang, Y. Two-dimensional hexagonal boron nitride for building next-generation energy-efficient devices. *ACS Energy Letters* 2021, 6, 985–996. doi:10.1021/acsenergylett.0c02427.
52. Xie, B., Li, C., Chen, J., Wang, N. Exfoliated 2D hexagonal boron nitride nanosheet stabilized stearic acid as composite phase change materials for thermal energy storage. *Solar Energy* 2020, 204, 624–634. doi:10.1016/j.solener.2020.05.004.
53. Wang, X., Maeda, K., Thomas, A., Takanabe, K., Xin, G., Carlsson, J.M., Domen, K., Antonietti, M. A metal-free polymeric photocatalyst for hydrogen production from water under visible light. *Nature Materials* 2009, 8, 76–80. doi:10.1038/nmat2317.
54. Meng, N., Ren, J., Liu, Y., Huang, Y., Petit, T., Zhang, B. Engineering oxygen-containing and amino groups into two-dimensional atomically-thin porous polymeric carbon nitrogen for enhanced photocatalytic hydrogen production. *Energy & Environmental Science* 2018, 11, 566–571. doi:10.1039/c7ee03592f.

55. Meng, W., Liu, X., Song, H., Xie, Y., Shi, X., Dargusch, M., Chen, Z.G., Tang, Z., Lu, S. Advances and challenges in 2D MXenes: From structures to energy storage and conversions. *Nano Today* 2021, 40, 101273. doi:10.1016/j.nantod.2021.101273.
56. Zhang, X., Zhang, Z., Zhou, Z. MXene-based materials for electrochemical energy storage. *Journal of Energy Chemistry* 2018, 27, 73–85. doi:10.1016/j.jechem.2017.08.004.
57. Chand, H., Kumar, A., Krishnan, V. Borophene and boron-based nanosheets: Recent advances in synthesis strategies and applications in the field of environment and energy. *Advanced Materials Interfaces* 2021, 8, 1–31. doi:10.1002/admi.202100045.
58. Kaneti, Y.V., Benu, D.P., Xu, X., Yuliarto, B., Yamauchi, Y., Golberg, D. Borophene: Two-dimensional boron monolayer: Synthesis, properties, and potential applications. *Chemical Reviews* 2022, 122, 1000–1051. doi:10.1021/acs.chemrev.1c00233.
59. Yadav, S., Sadique, M.A., Kaushik, A., Ranjan, P., Khan, R., Srivastava, A.K. Borophene as an emerging 2D flatland for biomedical applications: Current challenges and future prospects. *Journal of Materials Chemistry B* 2022, 10, 1146–1175. doi:10.1039/d1tb02277f.
60. Bai, Y., Liu, C., Shan, Y., Chen, T., Zhao, Y., Yu, C., Pang, H. Metal-organic frameworks nanocomposites with different dimensionalities for energy conversion and storage. *Advanced Energy Materials* 2022, 12, 1–52. doi:10.1002/aenm.202100346.
61. Elsayed, A., Elsayed, E., AL-Dadah, R., Mahmoud, S., Elshaer, A., Kaialy, W. Thermal energy storage using metal–organic framework materials. *Applied Energy* 2017, 186, 509–519. doi:10.1016/j.apenergy.2016.03.113.
62. Wang, H., Zhu, Q.L., Zou, R., Xu, Q. Metal-organic frameworks for energy applications. *Chem* 2017, 2, 52–80. doi:10.1016/j.chempr.2016.12.002.
63. Lv, J., Tan, Y.X., Xie, J., Yang, R., Yu, M., Sun, S., De Li, M., Yuan, D., Wang, Y. Direct solar-to-electrochemical energy storage in a functionalized covalent organic framework. *Angewandte Chemie International Edition* 2018, 57, 12716–12720. doi:10.1002/anie.201806596.
64. Yuan, H., Kong, L., Li, T., Zhang, Q. A review of transition metal chalcogenide/graphene nanocomposites for energy storage and conversion. *Chinese Chemical Letters* 2017, 28, 2180–2194. doi:10.1016/j.cclet.2017.11.038.
65. Yadav, S., Yashas, S.R., Shivaraju, H.P. Transitional metal chalcogenide nanostructures for remediation and energy: A review. *Environmental Chemistry Letters* 2021, 19, 3683–3700. doi:10.1007/s10311-021-01269-w.
66. Ramalingam, G., Kathirgamanathan, P., Ravi, G., Elangovan, T., Arjun Kumar, B., Manivannan, N., Kasinathan, K. Quantum confinement effect of 2D nanomaterials. *Quantum Dots: Fundamentals, Synthesis and Applications* 2020. doi:10.5772/intechopen.90140.
67. Ruan, S., Luo, D., Li, M., Wang, J., Ling, L., Yu, A., Chen, Z. Synthesis and functionalization of 2D nanomaterials for application in lithium-based energy storage systems. *Energy Storage Materials* 2021, 38, 200–230. doi:10.1016/j.ensm.2021.03.001.
68. Ming, X., Guo, A., Wang, G., Wang, X. Two-dimensional defective tungsten oxide nanosheets as high performance photo-absorbers for efficient solar steam generation. *Solar Energy Materials and Solar Cells* 2018, 185, 333–341. doi:10.1016/j.solmat.2018.05.049.

69. Wang, Z., Yu, K., Gong, S., Mao, H., Huang, R., Zhu, Z. Cu3BiS3/MXenes with excellent solar-thermal conversion for continuous and efficient seawater desalination. *ACS Applied Materials & Interfaces* 2021, 13, 16246–16258. doi:10.1021/acsami.0c22761.
70. Zhang, D., Hu, S., Liu, X., Chen, Y., Xia, Y., Wang, H., Wang, H., Ni, Y. Solar cells based on two-dimensional WTe2/PtXY (X, y = S, Se) heterostructures with high photoelectric conversion efficiency and low power consumption. *ACS Applied Energy Materials* 2021, 4, 357–364. doi:10.1021/acsaem.0c02283.
71. Zhang, J., Zhao, F., Zhang, Z., Chen, N., Qu, L. Dimension-tailored functional graphene structures for energy conversion and storage. *Nanoscale* 2013, 5, 3112–3126. doi:10.1039/c3nr00011g.
72. Lei, Z., Zhang, J., Zhang, L.L., Kumar, N.A., Zhao, X.S. Functionalization of chemically derived graphene for improving its electrocapacitive energy storage properties. *Energy & Environmental Science* 2016, 9, 1891–1930. doi:10.1039/c6ee00158k.
73. He, W., Zhou, L., Wang, M., Cao, Y., Chen, X., Hou, X. Structure development of carbon-based solar-driven water evaporation systems. *Science Bulletin* 2021, 66, 1472–1483. doi:10.1016/j.scib.2021.02.014.
74. Zhang, H., Ultrathin two-dimensional nanomaterials. *ACS Nano* 2015, 9, 9451–9469. doi:10.1021/acsnano.5b05040.
75. Bernardi, M., Palummo, M., Grossman, J.C. Extraordinary sunlight absorption and one nanometer thick photovoltaics using two-dimensional monolayer materials. *Nano Letters* 2013, 13, 3664–3670. doi:10.1021/nl401544y.
76. Liu, Y.L., Shi, Y., Yang, C.L. Two-dimensional MoSSe/g-GeC van der waals heterostructure as promising multifunctional system for solar energy conversion *Applied Surface Science* 2021, 545, 148952. doi:10.1016/j.apsusc.2021.148952.
77. Chandrasekhar, P.S., Dubey, A., Qiao, Q. (2020). High efficiency perovskite solar cells using nitrogen-doped graphene/ZnO nanorod composite as an electron transport layer. *Solar Energy*, 2020, 197, 78–83.

CHAPTER 3

Metal-Organic Frameworks for Energy Storage Applications

S.K. SUJA* and K. BERLIN ASHA

Department of Chemistry, Lady Doak College, Madurai, Tamil Nadu, India

*Corresponding author

ABSTRACT

Sustainable power sources like sunlight and wind are the alternative energy utilized by human beings. Due to unsteady yield of these fuel resources, innovations in electrochemical energy storage and conversion, such as rechargeable batteries, fuel cells, capacitors, and electrolyzers, are assuming importance toward productive and reasonable energy usage. These electrochemical devices contain components like the electrodes and electrolyte, where the charge storage and ions transport takes place. The performance of these devices is governed by the properties of the materials used in these components. Consequently, looking for better materials has been a significant journey for the advancement of future electrochemical energy-related innovations. Metal-organic frameworks (MOFs) are a group of rapidly developing permeable materials that have created much interest from specialists in different areas of research. The MOF's composition and porous structure is tuned by the starting materials and synthetic route providing flexible functionalities of MOFs. The high porosity and surface zone of MOFs are particularly reasonable for applications including capacity and association with guest species. These highlights trigger interests to investigate MOFs and related materials for electrochemical energy storage and conversion and are a fast-extending

Advances in Energy Materials: Materials Fabrication, Properties and Applications.
Lishin T, Maciej Jaroszewski, and Sabu Thomas (Eds.)

DOI: 10.1201/9781779643674-3

research field. This chapter explains the growth of MOFs and their utility toward renewable energy applications.

3.1 INTRODUCTION

3.1.1 ELECTROCHEMICAL ENERGY AND STORAGE

The conversion of chemical energy into electrical energy and vice versa is known as electrochemical energy, which includes redox reactions and electron transfer reactions. This requires electrochemical cells that can generate electrical energy from chemical reactions or to cause a chemical reaction, electrical energy is used by these cells [1]. Galvanic cell and electrolytic cell are the two types of cells used for electrochemical conversion.

Electrochemical energy storage is a method used to store electricity in the form of chemical constituents. In this storage technique, both electrical and chemical energy share the same carrier, the electron. Hence the loss of energy due to conversion is limited. Batteries, capacitors, and fuel cells are some common forms used for electrochemical storage and conversion.

3.1.2 NEED FOR ELECTROCHEMICAL ENERGY STORAGE SYSTEM

In this modern period, the need for energy grows without ceasing; on the other side, concerns about the depletion of resources and the problems associated with environmental pollution are becoming furthermore serious. Renewable and sustainable energy-storage technologies are the worldwide research in progress to arrive at a solution for the energy and environmental crisis. Renewable energy sources are advantageous and also have high applicability. But due to their irregular supply and difficulty in storing the energy that is harvested, many researches have focused in the development of effective and reliable energy storage devices [2]. Developments in the market like hybrid electric vehicles, plug-in hybrid vehicles, and full electric vehicles have raised the demand for the energy and power density of devices having electrical energy storage (EES). Batteries and supercapacitors are two significant and most hopeful advancements for EES attributable to their movability and compact size for on-request use [2]. They can be distinguished based on energy density, power density, efficiency, conductivity, voltage window, and current density (Figure 3.1). Despite their promising usage, there are limitations in the materials used in these devices which obstruct their rate of performance, safety, and life cycle. For example, polymeric materials and carbon materials with a conductive nature are moderately cheap, adaptable, and synthetically tuneable; however, they need physical and chemical solidity for device execution. On the other hand, strong inorganic materials, like metal oxides and silicon, are utilized as anode materials because of their powerful design and redox-dynamic locales. Nonetheless, slow ion dispersion of metal oxides limits charge/release rate capacities, and huge volumetric changes lead to mechanical [2] unsteadiness. Hence, materials with new, creative improvements to overcome the shortcomings of the existing are required. Disadvantages in these current stages rouse the disclosure and improvement of new materials for advanced energy storage devices [2, 3].

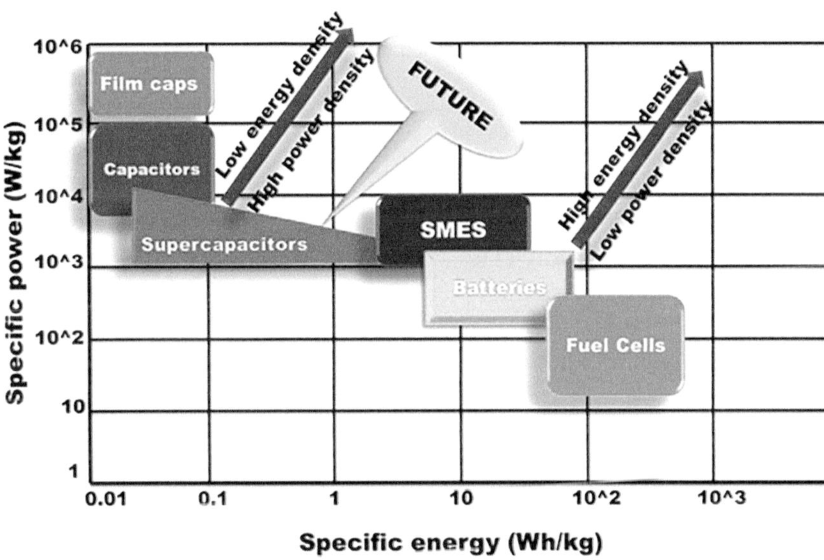

FIGURE 3.1 Ragone plot of distinct energy storage devices.

3.1.3 ELECTRODE MATERIALS USED IN ELECTROCHEMICAL STORAGE SYSTEMS

Portable electronics, transportation, and large-scale energy storage systems involve electrochemical energy storage technology which is critically important. The growing demand for energy storage devices with safety, long-term stability, low cost, high power densities, and high energy has created the necessity to design novel structures and high-performance electrode materials [4]. Porous graphene [5], doped materials [6], 1D porous materials [7], 3D ordered porous materials [8], mesoporous materials [9], nanostructured materials [10], covalent organic frameworks [11], carbon-based materials [12], polymer composites [13], metal hydrides [14], transition metal oxides [15], transition metal nitrides [16] are used as electrode materials in various EES systems. All these materials have a hierarchical structure, abundant low-cost resources, and show good electronic conductivity, specific capacity, stability, and security. For the past two decades, many research works have focused in designing metal-organic frameworks (MOFs) and MOF-derived materials for energy applications. Nearly 100,000 articles are published as research articles and patents in this field. This chapter deals with MOFs which is a class of porous materials used for the storage of electrochemical energy.

3.1.4 POROUS MATERIALS

Porous materials have narrow and tuneable porosity and surface area. Due to their flexibility in structure and functionalities, they find wide applications in catalysis, adsorption studies, sensing, separation and storage. These materials are broadly classified (Figure 3.2) based on their (1) pore size as microporous (pore diameter <2 nm), mesoporous (2–50 nm) and macroporous materials (>50 nm) and (b) framework building blocks as inorganic (silica, metal oxide, and metal phosphate), organic–inorganic hybrid (periodic mesoporous organosilica, MOF) and organic materials (porous carbon and organic porous polymers).

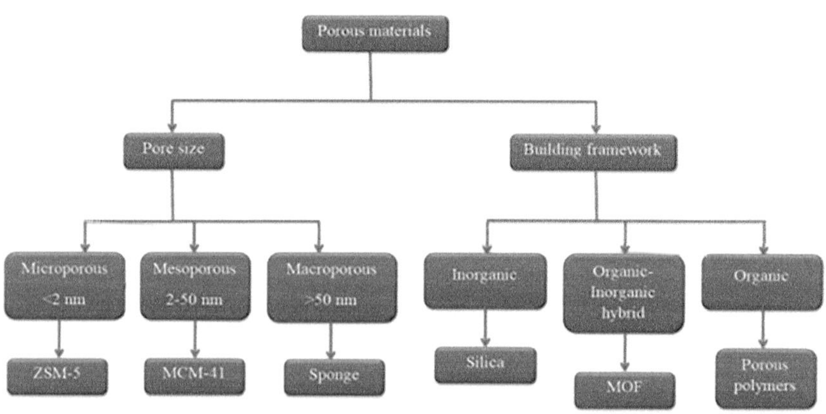

FIGURE 3.2 Classification of porous materials.

3.1.5 METAL-ORGANIC FRAMEWORKS

MOFs are porous coordination polymers formed by self-assembly of inorganic metals and multidentate organic ligands through coordination bonds as a crystalline network in a one-pot reaction. Usually, the synthesis of MOFs involves the mixing of metal salts and organic ligands in organic solvents at elevated temperatures to obtain the desired crystal growth [17]. MOFs are characterized by a crystalline arrangement of the atoms comprising the walls. MOFs exhibit significant advantages by providing structural diversity, flexibility, high porosity, large surface areas, and a wide range of pore sizes compared to other conventional mesoporous and nanoporous materials such as zeolites, carbon molecular sieves, and porous metal oxides [17]. Due to their unique properties, research on MOFs has become a vigorous and rapidly developing field in chemistry. Metals of the main group, transition series, inner transition group, and mixed metals with equal contribution are used for the synthesis of MOFs. Aromatic carboxylic acids and N-containing compounds are regularly utilized organic linkers in MOFs. In light of the size of inorganic structural components and ligands, the network of the framework, pores, and open networks with size from angstroms to nanometers can be available [18]. Their flexibility, long-range order, and rich host–guest interaction make MOFs ideal stages for distinguishing configuration highlights for advanced functional materials and precursors of derived materials for the storage of electrochemical energy and conversion advances (Figure 3.3) [2]. MOFs are used as electrodes, separators, and host material in batteries and supercapacitors.

3.1.6 MOF AND MOF-DERIVED MATERIALS FOR ELECTROCHEMICAL ENERGY STORAGE

As MOFs can be designed with the required properties with the use of suitable building blocks their properties can be channelized appropriately. This is of great advantage, and hence they are considered to be attractive candidates to fulfill the demands of advanced energy storage technologies. The synthetic flexibility attained through the combined effect of metal centers and linkers paves the way to overcome the limitations of materials used for energy storage. Due to the tuneable capabilities of MOF, functional materials having desired properties, unique morphology, and chemical composition can be synthesized. However, pristine MOFs tend to exhibit low conductivity and poor stability, which restricts their energy storage as well as conversion efficiency. Hence, to overcome this disadvantage, MOF-derived metal oxides are explored recently. Moreover, when MOF-derived metal oxide is combined with other materials to form the composite material, they show good performance. Hence, extensive research has progressed in this field.

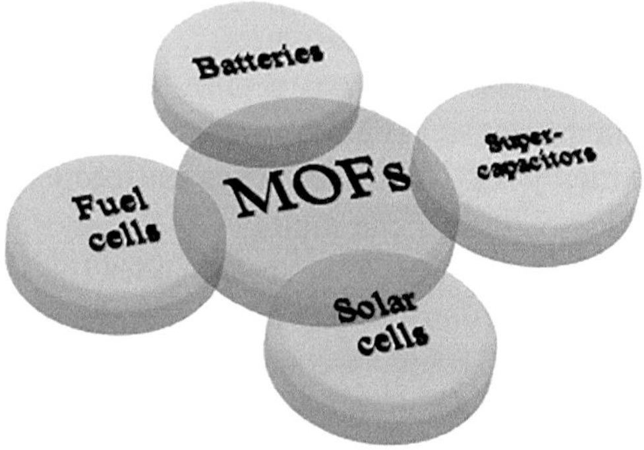

FIGURE 3.3 Metal-organic framework (MOFs) for electrochemical energy storage.

3.2 METAL-ORGANIC FRAMEWORKS IN ADVANCED BATTERY TECHNOLOGIES

The innovation of energy storage systems primarily depends upon the development of efficient electrode materials. An ideal electrode material should be highly porous, chemically/thermally stable, and possess excellent ionic/electrical conductivity. Different types of metal oxides and composites are used to enhance the efficiency of batteries. However, their poor conductivity and deformation upon charging/discharging result in low battery life and low efficiency. The synthesis protocols for carbon-based materials are too expensive. These are serious concerns in extending their applications to large-scale installations of energy storage devices. The fascinating characteristics of MOFs have been recognized based on the advent of their synthesis techniques. Hence, MOFs have been preferably employed in the development of the next-generation electrode materials. One can customize the MOF properties in line with the desired application by choosing appropriate metal ions and modulating the pore size using appropriate organic linkers.

The advantages of MOF-derived materials have been known for having synergistic properties while retaining the well-defined structure of the MOF [19].

3.2.1 LITHIUM-ION BATTERIES

Lithium-ion batteries (LIBs) have three essential components, namely, cathode, anode, and an electrolyte. The Li^+ ions deintercalate from the cathode upon charging and intercalate to the anode by moving through the electrolyte (Figure 3.4). The energy obtained during this process is stored and is released during the discharge process, where the Li^+ ions migrate toward the cathode. The electrode materials used as cathode and anode suffer from limited energy, power density, complicated synthetic process, and short life cycle [20]. In addition, the formation of Li dendrites due to an irregular transport of ions during cycling results in a low coulombic efficiency (Figure 3.5). Hence, novel materials to meet out these drawbacks are required.

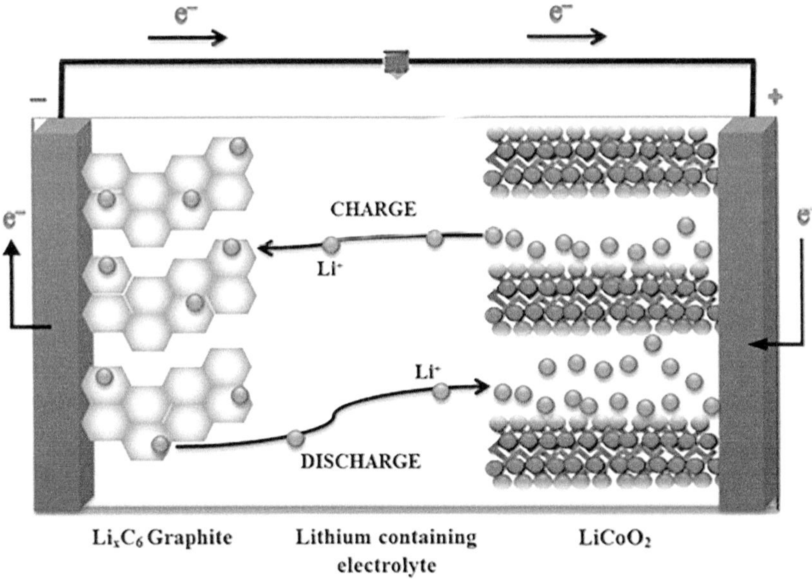

FIGURE 3.4 Working of Li-ion battery.

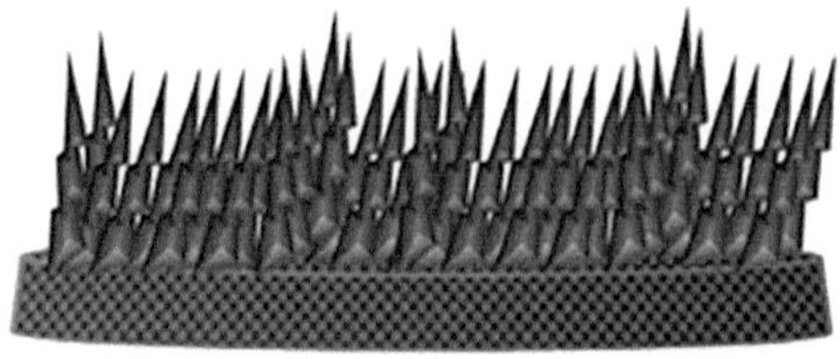

FIGURE 3.5 Formation of lithium dendrites on Li electrode.

3.2.1.1 MOFS FOR LIBS

MOFs with unique structure, highly porous, storage capacity, and high specific surface area serve as capable and potential electrode material for LIBs. The metal cations present in MOFs act as active sites for the redox reactions, and the crystalline open framework of MOF supports effective and reversible insertion/extraction of ions. The porosity of MOFs can lead to large space to accommodate lithium-ions and suitable lithium-ion pathway. MOFs and MOF-derived materials have served both as cathode and anode materials.

Physical and chemical stability, high electronic and ionic conductivity, excellent porosity, high surface area, and mass production potential are the favorable properties of materials (Figure 3.6) required for Li storage. Physical–chemical stability is the fundamental need for stable Li storage due to the fact that the active material in devices that experience volume expansion and contraction must preserve their structural integrity. The composition of linkers and metal nodes can be altered during synthesis to regulate physicochemical parameters. Therefore, it is crucial to choose a suitable synthesis medium for MOFs to increase the physical and chemical stability, a crucial quality of materials employed in lithium storage devices [21–24].

Another parameter that is essential for an electrochemical device is electron and ion transport. The fundamental requirement to achieve high power and high capacity in lithium batteries is the storage of Li content which is affected by the ionic conductivity. Generally, MOFs with M–O bond are poor conductors but by modifying the linkers and metals, their electronic conductivity is improved. For example, in the Fe^{2+} analog of MOF-74, the electronic conductivity is improved significantly by using sulfur-containing linkers so that the d-electrons loosely coupled to Fe^{2+} can

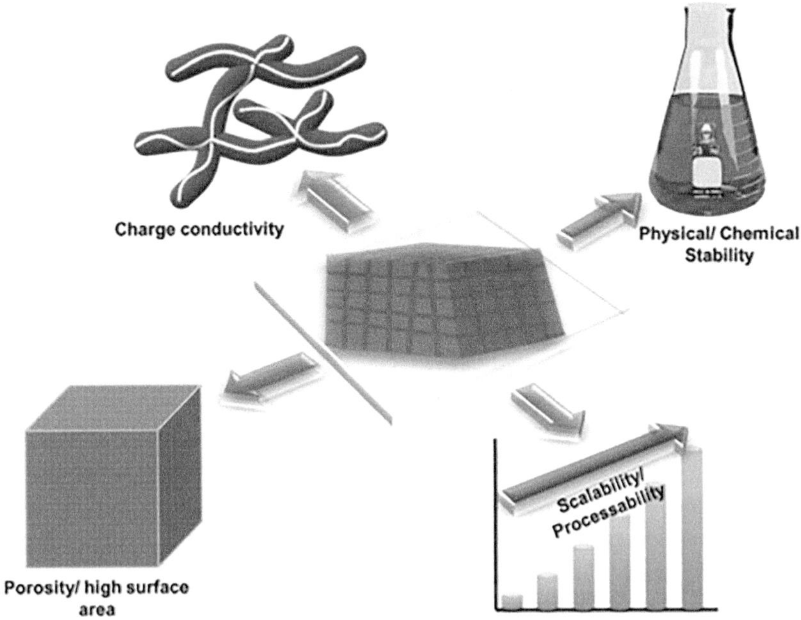

FIGURE 3.6 Favorable properties of metal-organic framework (MOF) for Li ion storage.

be carried by the soft sulfur linkers. Further, by having molecules that can allow electron hopping through π–π interactions, ionic guest molecules, the conducting property of MOF is improved. An ionic framework with a mobile counter ion can also be designed to significantly increase the ionic conductivity [25–31]

Porosity and high surface area are the other two important parameters essential for the storage and transport of electrochemically active species and for the catalytic processes. The most characteristic features of MOFs are their large surface area and excellent porosity. By effectively controlling the pore size, surface area, and chemical characteristics of MOFs, the structural and chemical factors that influence electrochemical processes can be controlled. For instance, when there is a combination of micropores and mesopores in the MOFs, they provide strategic channels and spaces that are advantageous for Li storage. The MOFs with high porosity enable the storage of polysulfide active species in LSBs and is also useful for devices that depend on the chemical transformation of small molecules, such as oxygen in LOBs [32, 33].

For any practical applications, the material must have the potential to be mass-produced. MOFs are produced in bulk by companies such as BASF, MOFWORX, MOF Technologies, NuMat, novoMOF, and MOF Apps [34–36] adopting green synthesis strategies. The various synthetic methodologies adopted for the mass production of MOFs are electrochemical, mechanochemical, microwave irradiation methods, and solution-phase layer-by-layer deposition. Electrochemical synthesis promotes uniform and fast growth of MOF and to achieve the chosen thickness, direct deposition is performed. The mechanochemical method involves physical mixing like grinding, ball milling, etc. of linker and metal ions without the application of heat or use of any solvent. This method has the advantages of mass production, reduced costs, and reduced chemical waste. Microwave irradiation requires less time and energy and allows rapid growth of MOFs from precursor solutions when compared with the conventional solvothermal method of synthesis. Solution-phase layer-by-layer deposition enables molecular and structural precision and control of film thickness [37–40].

3.2.1.1.1 MOFs as Cathode Material in LIBs

MOFs of various types have been used in LIBs as cathode materials. Fe(III)(OH)(1,4-BDC) also known as MIL-53(Fe), where MIL is Materials Institute Lavoisier, was used as electrode material in which Li^+ ions are inserted reversibly. During the electrochemical reaction, Fe^{3+} is converted to Fe^{2+} simultaneously Li^+ ions are embedded into the structure. The electrode discharge capacity, though not very prominent, has set a platform for exploring new materials for improvement. The material was modified with electro-active molecule say 1,4-benzoquinone and the electrochemical capacity was explored. 1,4-benzoquinone absorption increases electrochemical capacity and performs as a redox medium to increase electron transfer in MIL-53(Fe). In some MOFs, there are linkers like conjugated carboxylates, which facilitate efficient and reversible charge transfer via π–π interaction which is very vital for excellent Li storage. Another class of hybrid material is metal organophosphate open frame (MOPOF) which is composed of transition metal phosphate. For example, $K_{2.5}[(VO)_2(HPO_4)_{1.5}(PO_4)_{0.5}(C_2O_4)]$ a MOPOF showed the deintercalation of K^+ ions from the framework during charging/discharging cycles. Both Li^+ and K^+

ions intercalate and deintercalate, but in LIBs, during the charge/discharge cycle Li^+ ions migrate and pass through electrodes instead of K^+ ions. This is due to the ionic mobility, which is connected to the ionic radius of the ions. Here K^+ is heavier than Li^+ ions. The MOF containing Cu ion can also be used as cathode material. An example is Cu(2,7-AQDC) (2,7-H_2AQDC = 2,7-anthraquinonedicarboxylic acid), which had high battery recyclability. Apart from Li-ion battery cathode materials, there are few Li-ion performance cathode materials. MOF-74 (Ni), MIL-47, MIL-132, $Rb_xMn^{II}_y[Fe^{III}(CN)_6]\cdot nH_2O$, $Mn^{III}[Mn^{III}(CN)_6]$ are some examples for performance cathode materials [41].

3.2.1.1.2 MOFs as Anode Material in LIBs

Like cathode materials, there are MOFs used as anode material in LIBs. Cubic MOF-177 which is (Zn_4O(1,3,5-benzenetribenzoate)$_2$) has been used as anode and the charge/discharge capacities were observed. The reactions involved in MOF-177 electrode are as follows:

$$Zn_4O(BTB)_2\cdot(DEF)_m(H_2O)_n + e^- + Li^+ \rightarrow Zn + Li_2O$$

$$DEF + H_2O + e^- + Li^+ \rightarrow Li_2(DEF) + LiOH$$

$$Zn + Li^+ + e^- \leftrightarrow LiZn$$

Three other MOFs $Co_3(HCOO)_6$, $Zn_3(HCOO)_6$, and $Zn_{1.5}Co_{1.5}(HCOO)_6$ [42, 43] with diamondoid structure were used as anode and obtained stable capacities which demonstrated the improvements compared to MOF-177. The proposed mechanism was similar to the above.

For $Zn_3(HCOO)_6$:

$$Zn_3(HCOO)_6 + 6Li^+ + 6e^- \leftrightarrow 3Zn + 6HCOOLi$$

$$3Zn + 3Li^+ + 3e^- \leftrightarrow 3LiZn$$

For $Co_3(HCOO)_6$:

$$Co_3(HCOO)_6 + 6Li^+ + 6e^- \leftrightarrow 3Co + 6HCOOLi \text{ [42, 43].}$$

$Co_2(OH)_2$BDC (BDC=1,4-benzenedicarboxylate), Li_2(OBA) (H_2OBA = 4,4′-oxybisbenzoic acid), and Prussian blue (PB) analogs ($M_3^{II}[Co^{III}(CN)_6]_2\cdot nH_2O$ (M = Co, Mn)) are other MOFs used as anode materials.

Recently, a disc-shaped $Li_{4-x}K_xTi_5O_{12}$ material was derived from the MOF, MIL-125 (Ti) by hydrothermal process. MIL-125 (Ti) containing polymetallic sites were converted to disc-shaped $Li_{4-x}K_xTi_5O_{12}$ with good reversibility and cycle performance, retaining the porous structure. The product had excellent first discharge-specific capacities of 179.112 mAh g^{-1} at 1°C, and 178.011 mAh g^{-1} after 200 cycles, which indicated that it was an ideal anode material for lithium-ion batteries [44].

Bifunctionalized MOFs (BMOFs) have also been used for electrochemical studies. The pores being an integral part of BMOF frameworks, the Li-ions are easily stored in the BMOF during charging/discharging process. However, the low conductivity and distortion in the structure during cycling limit the application of MOFs as anode/cathode material. Further derived materials show improvement in the electrochemical properties used for cathode or anode [45]. The performance of few MOF and MOF-derived materials in terms of rate, reversibility, and recyclability is given in Figure 3.7.

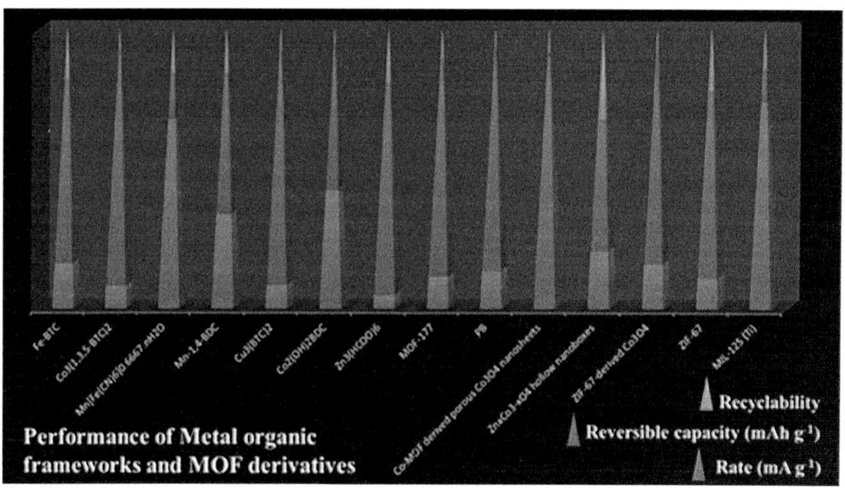

FIGURE 3.7 Performance of few metal-organic framework (MOF) and MOF derived materials.

3.2.1.2 *MOF-DERIVED OXIDES FOR LIBS*

For LIBs, MOF-derived oxides have a high specific potential and a long cycle life. Sintering PB analog ($Co_3[Co(CN)_6]_2$) yielded porous Co_3O_4

nanocages with a high and stable capacity. The small size, porous framework, and high surface area of porous Co_3O_4 nanocages allowed high electrochemical performance in LIBs. The nano building blocks of the Co_3O_4 cages facilitated the faster charge and mass transfer which enabled higher specific capacity than bulk materials. The hollow structures of the nanocages were able to shorten the length of Li^+ diffusion and allow complete contact with the electrolyte and hence were able to reduce the volume variation of the charge–discharge system. The largest specific surface area of the Co_3O_4 nanocages provided abundant active sites and also relieved the mechanical stress generated in the volume expansion/compression upon cycling [46]. Another method of preparing porous metal oxide nanosheets from MOF precursor is liquid phase deposition. Fe_2O_3 is another potential anode material vowing to its high theoretical capacity, nontoxicity, and low treatment cost. Fe_2O_3 microboxes obtained by annealing PB microcubes and spindle-like porous α-Fe_2O_3 from $Fe_3O(H_2O)_2Cl(BDC)_3 \cdot nH_2O$ (MIL-88-Fe) are examples for derived materials with high specific capacity [47].

Copper-based MOF-derived CuO also possesses various applications due to their abundance, cost-effectiveness, and eco-friendly nature. Pyrolysis of MOF-199 yielded nanostructured CuO which exhibited high initial capacity, good cycle performance, and excellent Columbic efficiency. Synthesis of MOFs-derived porous carbon matrices embedded with transition metal oxides enhances the electrochemical performance of the electrode material. Fe-MOFs were used as precursors to synthesize microcuboid-like carbon-decorated iron oxide (C-Fe_3O_4), which showed high reversible capacity and exceptional rate performance at different current densities. Controlled pyrolysis of isoreticular MOF-1 gave ZnO quantum dots with porous carbon that showed a high specific capacity and noticeable rate performance [41]. Another porous carbon material that was derived from MOF-5 was embedded with boron nitride (BN), in which B and N elements enhanced the electron transport speed and improved the conductivity simultaneously. By embedding BN lamellar system into the MOF-5-derived material, the surface area of the material was increased, and therefore it caused excellent cycling stability and high capacity characteristics in LIBs [48].

A three-dimensional P-doped graphene framework (PGF@pFe_2O_3NF) using MOF as the precursor of porous Fe_2O_3 proved to effectively promote lithium diffusion. This material has been used as an anode in LIB.

It exhibited excellent capacity (1106 mAh g^{-1}), rate performance (410 mAh g^{-1} at 4000 mA g^{-1}), and superlong cycling performance of 88.2% capacity retention [49]. An MOF-derived $NiCo_2O_4$ nanocubes combined with r-graphene oxide (GO) was used as electrode material for electrochemical energy storage and showed enhanced electrochemical performance. The surface of $NiCo_2O_4$ with nanoscale porosity provides a better cycling stability, good electrical conductivity, reduces the aggregation of active materials, and provides a larger electrode surface. r-GO sheets act as a conductive material in the nanostructure to provide the benefits of electrical double layer and to complement the action and performance of the redox reactions supported by the nanostructure. The $NiCo_2O_4$/r-GO-based electrode had an outstanding stability, with capacity losses only of ca. A total of 11% after very long operation periods (>2000 cycles) [50].

3.2.1.3 MOF-DERIVED SULFIDES FOR LIBS

The abundance of sulfides and its high theoretical specific capacity had made them potential electrode materials for LIBs. Novel porous sulfide/carbon composites accommodate and promote rapid diffusion of ions during the lithiation/delithiation process. The carbon material increases the electronic conductivity of the material. A graphene based composites (rGO)@CoSx and CoSx–rGO–CoSx were derived from MOFs/GO precursors and showed high specific capacities, rate capabilities, and long life cycle. Thermal carbonization of MOFs yields porous carbon materials with large surface area, distinct distribution of pore size, etc. As a result, very appealing methods for synthesising MOFs and MOF-derived materials electrode materials of next-generation LIBs is essential [41].

TABLE 3.1 Performance of Metal-Organic Frameworks (MOFs) and MOF-Derived Materials in LIBs

MOF	Electrode	Rate (mA g^{-1})	Reversible Capacity (mAh g^{-1})	Recyclability	References
MIL-53 (Fe)	Cathode	0.025 C	70	–	[41
MIL-101 (Fe)	Cathode	0.2 C	72	100	[41]
Cu(2,7-AQDC)	Cathode	1 mA s^{-1}	105	50	[41]
Cu-TCA	Cathode	2 C	45	200	[51]
Fe-BTC	Anode	30	120	35	[52]

TABLE 3.1 *(Continued)*

MOF	Electrode	Rate (mA g^{-1})	Reversible Capacity (mAh g^{-1})	Recyclability	References
$Co_3(1,3,5\text{-}BTC)_2$	Anode	100	1020	100	[53]
$Mn[Fe(CN)_6]_{0.6667} \cdot nH_2O$	Anode	2000	473	500	[54]
Mn-1,4-BDC	Anode	200	295	100	[55]
$Cu_3(BTC)_2$	Anode	100	974	100	[56
$Co_2(OH)_2BDC$	Anode	383	474	50	[42, 43]
$Zn_3(HCOO)_6$	Anode	50	750	200	[42, 43]
MOF-177	Anode	103	694	100	[42, 43]
PB	Anode	200	1192	100	[47,
Co-MOF-derived porous Co_3O_4 nanosheets	Anode	20000	300	2000	57]
PCFO-NC	Anode	2500	152	500	[58]
$CoFe_2O_4$ nanocubes					
$Zn_xCo_{3-x}O_4$ hollow nanoboxes	Anode	500	1141	800	[59]
ZIF-67-derived Co_3O_4	Anode	200	955	100	[60]
ZIF-67	Anode	200	1200	400	[61]
MIL-125 (Ti)	Anode	2000	310	400	[44]

3.2.2 LI-O_2 BATTERIES

The Li-O_2 batteries are made up of lithium as anode and oxygen as cathode. At the anode a lithium ion are released and passes through the electrolyte to the cathode. At the cathode, they react with oxygen to produce Li_2O and Li_2O_2. Li-O_2 batteries are striking due to their high energy density, which fulfil the requirements of future vehicle industry. Still, they face enormous challenges due to the following problems encountered.

1. The overall performance of the battery is affected by slow kinetics of oxygen reduction reaction (ORR) and oxygen evolution reaction (OER) which increases the overpotential.
2. The discharge product (Li_2O_2) which is insoluble and the side products (Li_2CO_3) deposited on the cathode inhibit the catalytic sites and diffusion pathway of oxygen and electrolyte, thereby creating overpotential.

To overcome this drawback, porous structure with high stability and activity must be explored.

3.2.2.1 MOFS FOR LI-O_2 BATTERIES

MOFs are good alternatives to improve the cycling performance of Li-O_2 batteries because of their high surface area, available metal sites, and flexible structure. MOFs can be used as selective gas sieves and storage media in Li-O_2 batteries. The metal nodes in MOFs serve as ORR/OER catalysts [2] (see Figure 3.8).

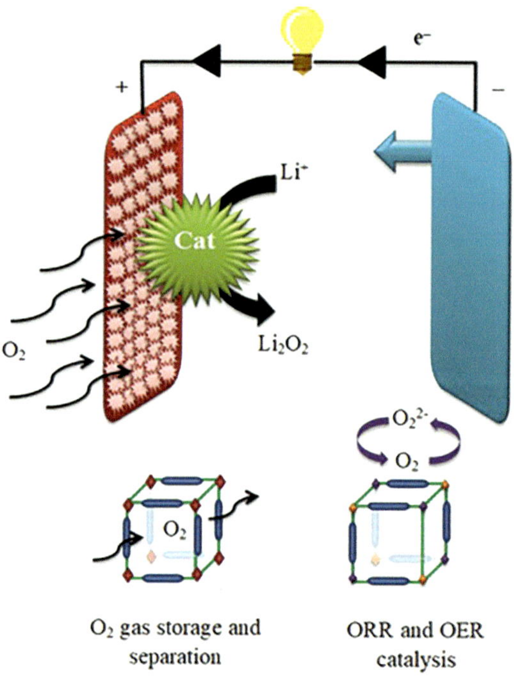

FIGURE 3.8 Li-O_2 battery for gas storage/separation and oxygen reduction reaction (ORR)/oxygen evolution reaction (OER) catalysis.

The Mn-MOF-74/Super P cathode showed a greater capacity compared to Super P cathode because of the open metal sites present in Mn-MOF-74 and high surface area, which increased the concentration of O_2 molecules in the micropores. Hence the material showed superiority in electrochemical performance for Li-O_2 batteries. Hollow and heterometallic MOFs have been used in catalytic reactions, especially involving gases. The hollow MOFs possess high surface area, well-organized catalytic active sites for gas absorption and storage, and possess large void space to fill with the discharge products [20, 62]. One such material is the hierarchical Zn/Ni-MOF-2 nanosheet-assembled hollow nanocubes which have served as a highly effective electrocatalyst for Li-O_2 batteries. By introducing structural defects in MOFs the catalytic performance can be improved.

Most of the studies on Li-O_2 batteries are conducted under oxygen environment. But CO_2 and moisture from the atmosphere poison the reactions and degrade the performance of the battery. Hence it is necessary to incorporate oxygen selective membrane. A novel mixed matrix membrane, that is, polydopamine-coated MOF crystals of CAU-1-NH_2 into a polymethylmethacrylate polymer has been used as a membrane in cathode, selective for oxygen in Li-O_2 batteries [20]. Bimetallic MOF-based composites are also proposed to be a promising cathode material for LOBs. One such example is Zn_xCo_y–C/CNT composites which were prepared in various Zn/Co ratios to control and adjust the physicochemical properties of the material. It was found that by controlling the structure and chemical composition of Zn_xCo_y particles, the robust cathode material was able to boost the electrochemical reactions as well as improve the reversibility of LOBs. An extremely high discharge capacity of ≈16,000 mAh g^{-1} for the Zn_4Co_1–C/CNT composite cathode and stable cycling performance up to 137 cycles was attained by improving the composition of the composite [59].

3.2.2.2 MOF-DERIVED MATERIALS FOR LI-O_2 BATTERIES

MOF-derived materials include oxides of metal, carbonaceous materials, metal oxide/carbon composites, and sulfides. They exhibit large surface area, hierarchical porous structure, and channels for mass transport. Porous spinel-type Co–Mn–O nanocubes, low-cost mesoporous two-dimensional cobalt sulfide nanosheets, nitrogen-doped graphene/graphene tube composites, hierarchical porous ZnO/$ZnFe_2O_4$/C (ZZFC) nanocages, spindle-shaped porous α-Fe_2O_3, ($Co_3(NDC)_3(DMF)_4$) are some examples of the materials derived from MOFs used in lithium-oxide batteries [20]. These materials have large pore volume, high specific capacity, conductivity, and high surface area.

TABLE 3.2 Performance of Metal-Organic Framework (MOFs) and MOF-Derived Materials in Li-O_2 Batteries

MOF	Electrode	Rate (mA g^{-1})	Reversible Capacity (mAh g^{-1})	Recyclability	References
Mn-MOF-74	Cathode	–	9420	–	[20, 62]
Fe(III)-MOF-5	Cathode	300	5000	15	[63
MIL-100 (Fe)	Cathode	0.1 mA cm^{-2}	600	30	[64]
MIL-100 (Cr)	Cathode	100	800	50	[65]
Co-MOF-74	Cathode	100	11,350 (primary capacity)	–	[66]
Zn_4Co_1–C/CNT composite	Cathode	–	16,000	137	[59]
Zn-MOF-5	Cathode	0.2 (mA cm^{-2})	1437 (initial discharge capacity)	–	[67]

3.2.3 LI-S BATTERIES

Lithium-sulfur batteries have found new scope for next-generation storage devices due to the low cost of sulfur and high theoretical specific energy. The difference between Li-S batteries and Li-ion batteries is the multielectron conversion of S to polysulfides, which results in the unusual specific capacity of Li-S batteries. The reactions occurring in Li-S batteries are as follows [68]:

$$S_8 + 2e^- \rightarrow S_8^{2-} \tag{3.1}$$

$$S_8^{2-} \leftrightarrow S_6^{2-} + \tfrac{1}{4} S_8 \tag{3.2}$$

$$2S_6^{2-} + 2e^- \leftrightarrow 3S_4^{2-} \tag{3.3}$$

$$3S_4^{2-} + 2e^- \leftrightarrow 4S_3^{2-} \tag{3.4}$$

$$2S_3^{2-} + 6Li^+ + 2e^- \leftrightarrow 3Li_2S_2 \tag{3.5}$$

$$Li_2S_2 + 2Li^+ + 2e^- \leftrightarrow 2Li_2S \tag{3.6}$$

During the discharge process, S_8 reduces to Li_2S_n, where, n = 4–8, and finally results in Li_2S_2 and Li_2S. This polysulfide shuttle leaches out polysulfides from the cathode and inactivates the electrode surface during cycling thereby limiting the cyclability of Li-S batteries (Figure 3.9). Moreover, it leads to a high self-discharge rate, large internal resistance, severe capacity fading, poor life cycle, and is unsafe. To overcome all these disputes, a host material is required that encapsulates sulfur within its matrix and improves conductivity. Lithium and sulfide ions are stored in Li-S batteries by host–guest interactions. Functional groups present in the linkers can also be used to capture polysulfides and decrease capacity loss (Figure 3.10). MOFs having high surface area and porosity can be used for the modification of sulfur cathode.

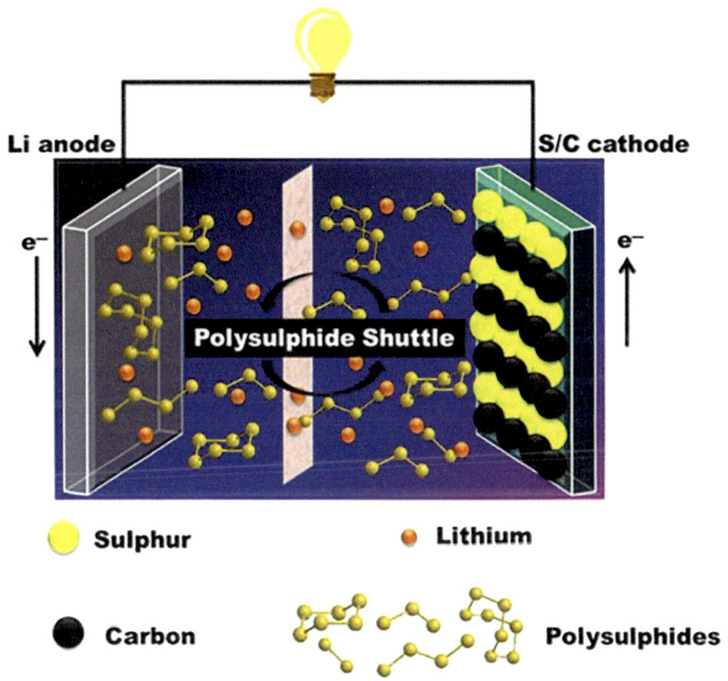

FIGURE 3.9 Polysulfide shuttle in Li-S battery.

3.2.3.1 MOFS FOR LI-S BATTERIES

The use of MOFs as host for sulfur in Li-S batteries was first proposed in the year 2011. MIL-100(Cr) with a high surface area produced a remarkable increase in the retention of the capacity of Li-S batteries [20]. Successively, many research works focussed on using MOFs to embed S into its pores. MOFs like HKUST-1, $Ni_6(BTB)_4(BP)_3$ (BTB = benzene-1,3,5-tribenzoate and BP = 4,4′-bipyridyl) [20], have suitable pore space and open M^{2+} (M = Cu^{2+}, Ni^{2+}) sites to confine sulfur into them and show high capacity and stable life cycle. In the case of Ni-MOF, the presence of both mesopores and micropores and strong interactions between the Lewis acid Ni^{2+} and polysulfide which acts as a base, inhibit the diffusion of polysulfides out of the pores. As an advancement, hybrid nanocomposites as sulfur hosts were synthesized to enhance the electrochemical performance of Li-S batteries [20]. For example, MIL101(Cr)@rGO composite was used in Li-S batteries. MIL101(Cr) provided cavities of microporous and mesoporous nature while the electronic conductivity was improved by the graphene sheets in MIL101(Cr). Sulfur was incorporated into the MIL101(Cr)@rGO composite to promote electron transfer and also to slow down polysulfide dissolution. This prepared composite cathode presented high electron conductivity and also had a good ability to capture lithium polysulfides [69]. Hence MOFs can be used as hosts to capture sulfur and to improve the capacity and life time of Li-S batteries.

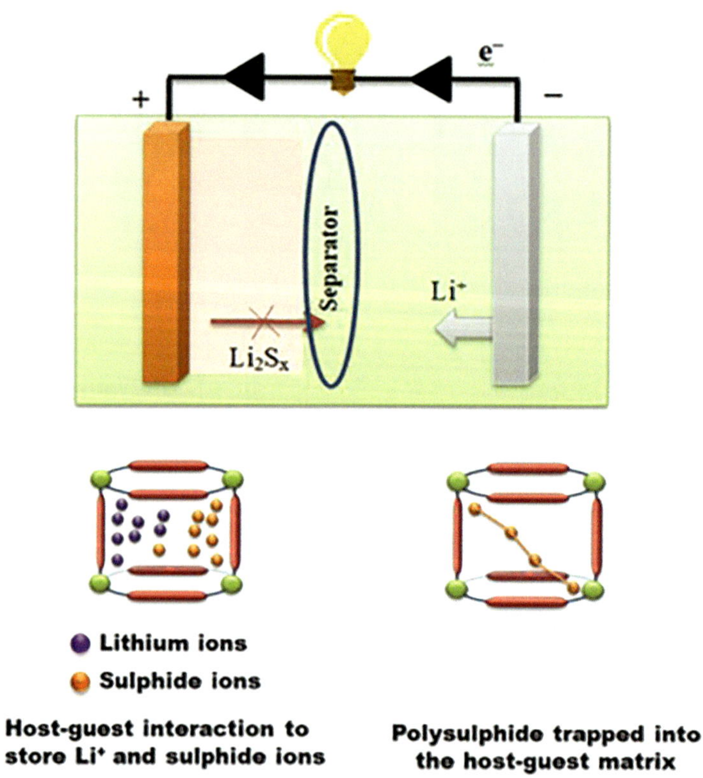

FIGURE 3.10 Metal-organic framework (MOF) as separator to control the polysulfide shuttle in Li-S battery.

Functionalizing the separator can be one solution to overcome the issue of polysulfide intermediates shuttle and the dendrite growth on the Li-metal surface lithium-sulfur (Li-S) batteries. A multifunctional MIL-125(Ti)-modified polypropylene/polyethylene material has been used as a separator for advanced Li-S batteries. The Ti-containing MOF features an open-skeleton construction, a high intrinsic microporosity, Lewis acid characteristics, better electrolyte wettability, and less resistance compared to pristine MOF. The distinct coating acts as a region of barrier to seize the polysulfides without affecting the Li^+ transport. The formation of lithium dendrites is inhibited by the highly ordered micropores in the MOFs, enabling uniform Li^+ plating, thereby allowing the separator to perform with high stability and rate capability. The capacity retention of over 60% upon 200 cycles is realized [70].

Recently, CSUST-1 a cerium MOF containing Ce(III, IV) ions with many oxygen vacancies (OVs) has been synthesized and the CSUST-1/CNT composite has been used as a separator coating material in the Li-S battery. The open Ce sites in Ce(III,IV) arrays and the OVs function synergistically to speed up the polysulfide redox kinetics as well as the transportation of Li^+ ions. Subsequently, the Li-S battery coated with CSUST-1/CNT separator showed high specific capacity (1468 mA h/g), long-term stability (1200 cycles), and a very low rate of decay (0.037%) per cycle. The separator demonstrated excellent performance even at high sulfur loading (8 mg/cm^2) [71].

A Co_9S_8 nanoflower cluster material was used as a functional separator for a high-performance Li-S battery. The polysulfide shuttle was suppressed by the Co_9S_8 cluster, which effectively interacted with the polysulfides. Hence the electrochemical performance of Li-S batteries can be enhanced by constructing effective polysulfide shuttle barriers like Co_9S_8 [72].

3.2.3.2 MOF-DERIVED CARBON-BASED MATERIALS FOR LI-S BATTERIES

MOFs are pyrolyzed under an inert atmosphere to produce carbonized products with micro-, meso- and macropores, high specific area, and pore volume. They serve as host materials to trap sulfur during the diffusion of polysulfides by channel adsorption. The heteroatom-doped porous carbon nanoplates (HPCN) have hierarchical structures derived from MOF-5 showed high surface area and large pore volume. When sulfur was embedded in HPCN the cycling performance and the specific capacity increased.

The electron transport was facilitated by the conductive carbon framework in HPCN and the diffusion of polysulfides was suppressed by the porous structure of HPCN. Additionally, a fast diffusion of Li^+ was enabled by the remaining pores in the composite. A study on the performance of the cathode material with respect to the pore size showed that carbon hosts with mesopore volume 2–50 nm exhibited large initial discharge capacity while the hosts with micropore volume of <2 nm had better life cycles. Another nitrogen-doped carbon with microporous and mesoporous structures from ZIF-8 was able to effectively confine sulfur into its structure. As nitrogen doping increases the interaction between carbon and sulfur increases thereby increasing the electrochemical stability. A novel 1D French fries-like porous carbon structure facilitates the electron/lithium ion transport and avoids the loss of sulfur into the electrolytic solution. The cathode showed high discharge capacity even after 100 cycles.

In addition, MOF-derived metal oxides can improve the electrochemical performance by trapping lithium polysulfides through polar surface interactions or Lewis acid–base interactions which improves the electrochemical performance of Li-S batteries. The role of MOF-derived metal oxides is to chemically adsorb lithium polysulfide and also to confine lithium polysulfide by physical adsorption, and finally to provide volume change space for sulfur during charging/discharging process.

TABLE 3.3 Performance of Metal-Organic Framework (MOFs) and MOF-Derived Materials in Li-S Batteries

MOF	Cathode/ Separator	Rate ($mA\ g^{-1}$)	Reversible Capacity ($mA\ h\ g^{-1}$)	Recyclability	References
MIL-100 (Cr)	Cathode	–	–	–	[65]
HKUST-1	Cathode	0.5 C	286	300	[20]
Ni based MOF ($Ni_6(BTB)_4(BP)_3$)	Cathode	0.1 C	611	100	[20]
ZIF-8	Cathode	0.5 C	950	250	[20]
ppy-S in PCN-224	Cathode	10 C	670	200	[73]
$Na_2Fe[Fe(CN)_6]$	Cathode	1 C	770	100	[74]
ZIF-67	Cathode	300	949	300	[75]
LPS-UiO-66	Cathode	100	375	600	[76]
HKUST-1@GO	Separator	1 C	855	1500	[77]
ZIF-8	Separator	2.5 C	403	300	[20]

TABLE 3.3 *(Continued)*

MOF	Cathode/ Separator	Rate ($mA\ g^{-1}$)	Reversible Capacity ($mA\ h\ g^{-1}$)	Recyclability	References
$Ni_3(HITP)_2$	Separator	0.1 C	1220	300	[78]
UiO-66/Nafion	Separator	0.1 C	1056	200	[79]
Ce-MOF-2 (Ce-MOF-808)/CNT	Separator	0.1 C	838	800	[80]
MIL-125(Ti)	Separator	0.2 C	1218	200	[70]
CSUST-1	Separator	2 C	538	1200	[71]
Co_9S_8 nanostructure-derived MOF	Separator	1 C	500	200	[72]

3.2.4 LI-SE BATTERIES

Selenium, which has similar chemical properties as that of sulfur finds application as electrode material in batteries. The lithiation/delithiation process is similar to sulfur. Selenium being a metalloid, has higher electronic conductivity is higher than S, hence enabling fast electrochemical reaction during the lithiation/delithiation process.

$$2Li^+ + Se + 2e^- \leftrightarrow Li_2Se$$

In spite of these advantages, Li-Se batteries also encountered with several drawbacks. Upon lithiation there is huge volume expansion at the Se cathode due to the lattice structure lacking in empty space. Also, Se cathode suffers from low Coulombic efficiency and cyclic performance due to the formation of polyselenides in solution, which leads to low utilization of Se and the capacity also fades rapidly. Hence to avoid the dissolution of polyselenides and to increase the void space, Se is embedded with porous carbon materials derived from MOFs.

3.2.4.1 MOFS FOR LI-SE BATTERIES

The porous carbon–selenium composite stops the formation of polyselenides, provides more space and also increases the electronic conductivity. MOF-5-derived Meso-C@Se, Se-ZIF-8-derived microporous carbon polyhedra (Se-MICP) [20], Al-MOF-derived nitrogen-doped layered carbon sponge (NCS)/Se-50 composite are few examples of Se combined with porous carbon material derived from various MOFs.

The mesoporous and microporous carbon has high electrical conductivity, high specific area, and large pore volume, which can confine high amount of Se, stop the diffusion of polyselenides, and thereby reduces the volume expansion during the discharge process [20]. Similarly, NCS has lavish micropores and mesopores that are interconnected and connected between layers, shows high rate capability and cycle performance. The bulk Se is confined in the micropores of NCS under vacuum and high temperature thereby reducing its dissolution during cycling. The mesopores absorb the electrolyte and allows quick transfer of Li^+ ions. The NCS exhibit good electric conductivity. The presence of heteroatom facilitates charge transfer, shows better performance and also improves the interaction between C and Se.

3.2.5 SODIUM-ION BATTERIES

Sodium resources are of low cost and widely distributed. The components and working principle of sodium-ion batteries (SIBs) are comparable with LIBs. Hence they can be used as promising competitor for next-generation EES (Figure 3.11). By analogy, the materials used for LIBs can also be explored for SIBs. But due to the larger size of sodium ions compared with lithium ions, Na^+ diffusion requires a larger tunnel size. Therefore the materials show poor kinetics of Na^+ ion insertion/extraction.

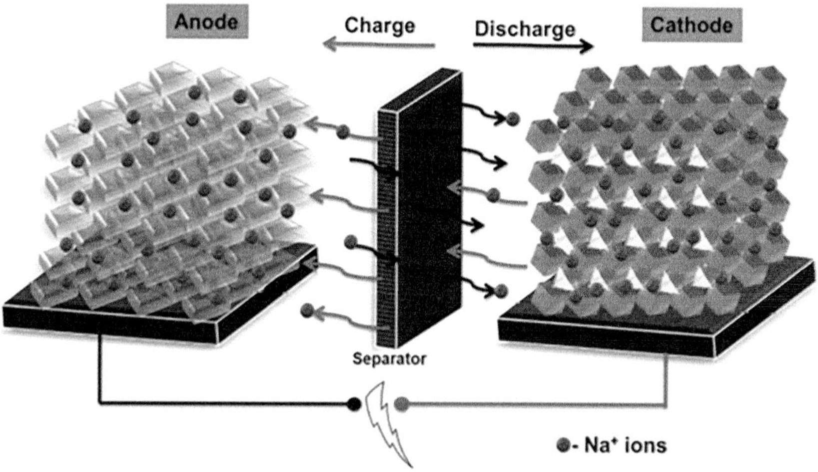

FIGURE 3.11 Sodium ion battery.

3.2.5.1 MOFS FOR SIBS

MOFs obtained by linking metal ion and flexible organic ligands are considered to be efficient electrode materials for SIBs. The PB analog (PBA)-nickel hexacyanoferrate nanoparticle was used for Na^+ insertion/extraction. The electrode possessed tremendous rate capability, high energy efficiency, and long life cycle [20]. The capacity did not fade over 5000 cycles in aqueous sodium electrolyte. Similarly, PBA-copper hexacyanoferrate also showed ultralong life cycle and high retention of capacity. The sodium storage property of NaCoHCF ($Na_xCoFe(CN)_6$) nanocrystals in aqueous electrolyte was studied and found to have high capacity due to its crystal and lattice integrity and also showed extraordinary capacity retention. In case of energy storage at a larger scale, PBA are exceptional candidates due to their remarkable performance, usage of low-cost electrolyte, and scalable production. PBAs such as ($KMFe(CN)_6$, M = Fe, Mn, Ni, Cu, Co and Zn) [81], and $Na_xMnFe(CN)_6$ serves as new cathode material for SIBs in the presence of saturated $NaClO_4$ in 1:1 ethylene carbonate/diethyl carbonate (EC/DEC) electrolyte. The development of new ion intercalation materials are of high interest owing to their applications toward SIBs [81].

3.2.5.2 MOF-DERIVED METAL OXIDES FOR SIBS

A classic treatment to improve the reaction kinetics, electronic conductivity and to make adjustments due to any volume change is using porous nanostructures and carbon modifications. Transition metal oxides such as Fe_2O_3, Co_3O_4, and CuO prepared from MOF precursors have been used as anodes for SIBs. A Ni-MOF was used as precursor from which a hollow nanostructure NiO/Ni/grapheme was synthesized. This material was used as an anode in SIBs and showed good rate of performance and stable life cycle. The hierarchical structure was able to adjust the volume change of active materials during consecutive cycles. The graphene layer facilitated fast electron transfer as well as promoted the formation of solid electrolyte interface film [20].

3.2.5.3 MOF-DERIVED CARBON MATERIALS FOR SIBS

Mesoporous carbon materials with narrow pore size distribution are derived from MOFs by nanocasting and pyrolysis. They significantly enhance the performance and kinetics of sodium ion storage. Pyrolysis of ZIF-8 produced ZIF-C with pore size <1 nm. The material showed high capacity, and reduced the irreversible capacity loss in the initial stage, and supported the reversible storage of sodium ions later. By thoroughly studying the influence of porous structure, pore volume, heteroatom doping [20], and specific areas on the storage of sodium ions excellent materials could be designed for further development of SIBs.

Recently, ZIF-derived porous $ZnxCo_{3-x}O_4$ hollow nanoboxes have been applied for LIBs and SIBs. The porous structure and hollow interspace provided valuable channels and voids for lithium/sodium ion transfer and storage. As an anode for LIBs, the electrode delivered a reversible capacity of 1141.7 mAh g^{-1} with a high coulombic efficiency of 95.6% at 100 mA g^{-1}, while the capacity was obtained at 625 mAh g^{-1} after 800 cycles at 500 mA g^{-1}. Applying a high current of 3.0 A g^{-1}, the specific capacity can still be maintained at 484 mAh g^{-1}. When used for SIBs anode, the reversible capacity was remained at 310 mAh g^{-1} with capacity retention of 90.4% even after 100 cycles [82].

3.3 METAL-ORGANIC FRAMEWORKS IN SUPERCAPACITORS

Another energy storage device is the supercapacitor, which holds high-power energy and a long life cycle. The components of a supercapacitor are shown in Figure 3.12. Electrical double-layer capacitors (EDLCs) and pseudocapacitors are two main types of supercapacitors. In EDLCs, when voltage is provided there is release of ions at the electrode-electrolyte interfaces. The energy produced is stored at the electrode surface during the electrostatic separation of charged ion-pairs whereas in battery the energy is stored during electrochemical conversion, which is a limitation of redox kinetics. Hence the lifetime, specific power, cycling efficiency, charging, and discharging in supercapacitors are superior to batteries. In EDLCs, carbon materials with high porosity and surface area are used. But the charge storage and energy densities of EDLCs are very low. Hence it is essential to replace them with pseudocapacitive materials that are electrochemically active and with higher energy density. Transition metal oxides and sulfides can be used for that purpose.

TABLE 3.4 Summary of the Types of Batteries, a Technology Overview According to the Most Important Aspect of Performance of the Battery and the Advantages of Metal-Organic Framework in Batteries

Type of Battery	Technology Overview	MOF in Batteries—Advantages and Uses	Performance Focussed
LIB	1. The intercalation/deintercalation of Lithium ions (Li^+) between the cathode and anode happens internally. The electron motion occurs in the opposite direction externally in the circuit. This migration helps the battery to power the device. 2. They have low theoretical capacity and limited rate capability.	• MOFs are used as advanced cathode and anode materials that offer higher Li storage capability than those of commercial materials. • The crystal structures of MOFs can host a large amount of Li^+ with excellent structural stability. • The porous structure of MOF is beneficial for boosting interfacial reaction kinetics. It is also crucial for the improvement of dimensional stability, superior rate capability, and cycling stability. • MOFs have been utilized as functional materials for separators, directly deposited on the polymeric membranes, providing various functionalities such as high permeability, high strength, and high heat resistance for improving the performance and safety of LIBs	Stability
$Li\text{-}O_2$	1. Oxygen is used as the cathode material and lithium is used as the anode material. 2. Energy density equivalent to 5–10 times that of LIBs can be realized.	• MOFs are used as cathode materials owing to the large surface area and pore volumes. • The charge transfer reactions occur on the surface of MOFs, and discharge products can be accommodated in the free spaces of MOFs. • MOFs improve the reversibility of LOBs by lowering the over-potential that arises from the discharge products that are decomposed while charging.	High energy
Li-S	1. Using sulfur as a cathode material and lithium as an anode material. 2. Possible to realize an energy density three times higher than that of a LIB. 3. Eco-friendly and low cost because uses sulfur waste, a by-product of crude oil refining.	• The large surface area and porosity which can be varied are the merits of MOFs as cathode materials for LSBs. • The solid sulfur and polysulfide molecules can be effectively confined in the free-space of MOFs, resulting in reduced side reactions. • MOFs are also used as absorbents for polysulfide molecules on the separator, leading to the improved reversibility of LSBs.	High energy and low cost of resource

TABLE 3.4 *(Continued)*

Type of Battery	Technology Overview	MOF in Batteries—Advantages and Uses	Performance Focussed
Li-Se	Using selenium as a cathode material and lithium as an anode material. Selenium being a metalloid, its electronic conductivity is higher than S Possible to realize an energy density higher than that of a LIB.	• MOF-derived materials are used as cathode and separators in Li-Se batteries. • Rapid self-discharge, poor cycling stability and low coulombic efficiency from polyselenide shuttling in Li-Se batteries are overcome by MOFs.	High energy
SIB	1. A battery that uses sodium ions as ion carriers instead of lithium ions. 2. Low-cost battery production is possible by replacing lithium, which is vulnerable to price and supply stability. 3. Some serious issues in SIBs are their poor lifetime and low discharge capacity.	• MOFs are promising candidates as anode material, as they provide rapid charging and good stability at very high temperatures. • MOF-derived materials show high electrical conductivity.	Low cost and abundance of resource

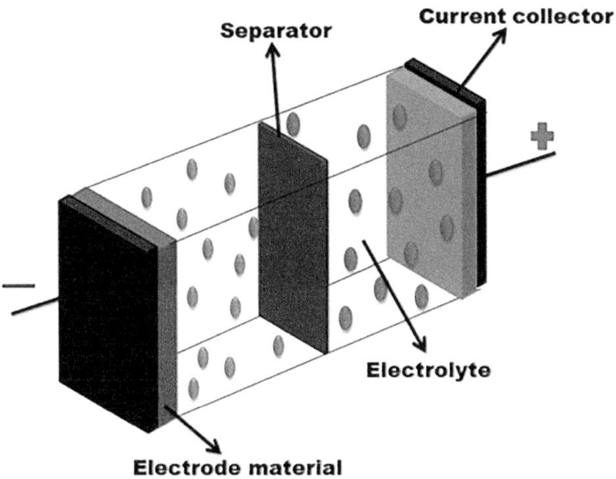

FIGURE 3.12 Components of a supercapacitor.

The electrochemical performance evaluation criteria for supercapacitors mainly include energy density, power density, and cycling lifespan. The energy density (E) is used to estimate how much energy can be stored. Utilizing power density allows one to assess the amount of power delivered per unit mass. Cycling stability is the key quality factor of a supercapacitor electrode and it is a measurement of the capacitance retention after a given working time or charging/discharging cycle number. A material with excellent capacitance and rate capability but suffering with a limited lifetime then it cannot be practically applicable. The widely accepted standard for the end-of-life of a supercapacitor is the time when 20% of the capacitance is lost or the internal resistance is doubled. The percentage of capacitance retained from that of the initial value is termed capacitance retention and it is usually applied to evaluate the cycling stability of the electrode materials.

3.3.1 MOFS FOR SUPERCAPACITORS

Nowadays MOFs and MOF-derived materials have gained more interest due to their high surface area, flexible structure, large diversity of metal ions, and controllable pore size. For EDLC, pseudocapacitive applications, MOFs serve as templates and precursors for preparing metal oxides/sulfides and porous carbon, which act as electrodes [83]. The power density, energy density, and cycling stability of the supercapacitors are influenced by the characteristic features of the electrode materials, like the morphology, porosity, and size which control the cell potential and capacitance. Extensive study on various electrode materials and their functions based on the structure and composition has been explored. MOFs are carefully designed as electrode materials to provide larger surface area, and active sites to increase the electrode/electrolyte contact area and to exhibit high performance. Hence the electrode material can be designed or prepared with a particular morphology and structure by deriving and transforming MOF precursors thereby improving the electrochemical performance [84].

3.3.2 MOFS AS ELECTRODES FOR SUPERCAPACITORS

MOFs with varying metal centers and organic linkers have been used as electrodes for EDLCs. One of its kinds is Co8-MOF-5, a mixed metal-based MOF that possesses good electrical conductivity but low specific capacitance. Hence tuning of the material was required to increase the specific capacitance value. By changing the metals or the organic linkers and using electrolytes that give rise to pseudocapacitance, this could be achieved. Another Co-based MOF with three organic ligands differing in their carbon chain length tuned the pore size thereby modifying the surface area and capacitive properties. When a longer organic linker was used, the porosity and surface area were increased and there was enhancement in capacitance. The high porosity and host–guest interaction of MOFs is beneficial for the storage of charged species in supercapacitors (Figure 3.13). The electrochemical performance of the material was attributed to the interconnected microstructure with few interfaces in structure.

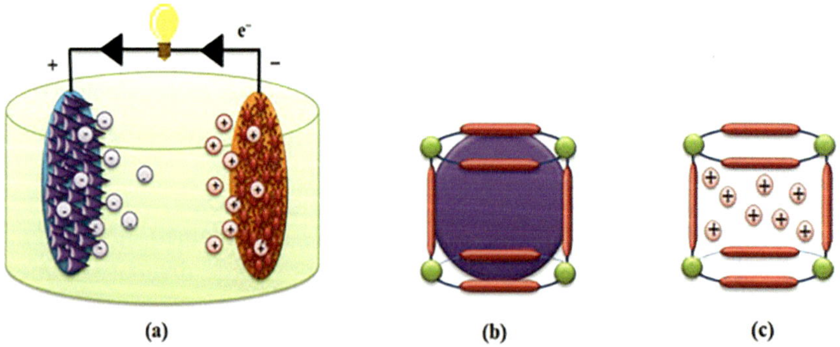

FIGURE 3.13 (a) Separation of charges in a supercapacitor, (b) enhanced surface area to accommodate guest molecules/ions, (c) storage of charged species in metal-organic framework (MOF).

To improve the specific capacitance and stability of the electrode materials used in supercapacitors, the synthetic strategy and structural modification are required. A Ni-based MOF with a layered structure has been used as electrode material in supercapacitors. For example, UiO-66 a Zr-based MOF had been synthesized under different temperature conditions which resulted in MOF with varied particle sizes and degrees of crystallization. The MOF obtained at low temperature had the smallest particle size and showed the highest specific capacitance. The capacitance of a series of MOFs such as MOF-5, nNiMOF-74, nHKUST-1, nMOF-177, nUiO-66, and nMOF-867 with a variety of structures was compared and it was found that nMOF-867, a Zr-MOF exhibited large capacitance per unit area than other MOFs.

Though there are few MOFs with good supercapacitive properties, due to their low conductivity, their potential application to be a supercapacitor was not effective causing poor rate of performance and fading of capacitance over cycles. Hence an effective scheme was put forward. By interweaving ZIF-67 MOF crystals with a conductive polyaniline chains, the conductivity was improved. Further, a flexible solid-state supercapacitor with electrode material such as PANI-ZIF-67-CC was developed and was used to power LEDs. These MOF-based solid-state supercapacitors show great potential for flexible and wearable electronics.

The use of rGO along with MOF has gained more attraction due to their high electrical conductivity, electrochemical stability, and surface area. For instance, the super capacitance behavior of Ni-doped MOF-5 with rGO was investigated. Nickel present in Ni-doped MOF-5 undergoes reversible redox reaction which involves two electron shuttles between the metal, Ni, and $Ni(OH)_2$ in alkaline electrolyte. The charge transfer resistance was reduced by rGO leading to increased capacitance than contributed from Ni doped MOF-5 alone. Similarly, Ni-MOFs grown on CNT surface presented high energy density and exceptional life cycle.

3.3.3 MOF-DERIVED METALLIC OXIDES/SULFIDES FOR SUPERCAPACITORS

MOF-derived materials, such as metal oxides and carbon-based composites are prepared by decomposing MOFs at appropriate temperatures and are used for supercapacitors. CeO_2, NiO_2, $ZnCo_2O_4$ nanoparticles, hollow Co_3O_4, $Ni_xCo_{3-x}O_4$ nanoparticles, nitrogen-doped carbon/Mn_3O_4 composites, Fe_3O_4/carbon composites, Cu–Cu_2O–CuO/C composites are some MOF-derived materials used for supercapacitors. Due to their excellent thermal properties and reactivity, MOF-derived materials are also used as templates for developing advanced porous functional materials. For example, Ni-Co PBA nanocubes by anion exchange method were transformed into cubic NiS nanoframe which exhibited improved electrochemical properties for electrochemical capacitors. Other examples of MOF-derived materials exhibiting electrochemical properties for capacitors include porous $Cu_{1.96}S$/C polyhedra from HKUST-1, sulfur powders, and Cr_2O_3 nanoribbons/carbon composite and cobalt sulfide polyhedral nanocages derived from ZIF-67 nanocrystals [20].

MOF-derived materials such as graphene-based composites and binary oxides also show excellent electrochemical properties for high-performance supercapacitor applications [20]. For example, porous Co_3O_4/$ZnFe_2O_4$ hollow nanocomposites with starfish shape are prepared from $Co_3[Co(CN)_6]_2 \cdot nH_2O$ and Fe_{III}-MOF-5 (Fe_{III}-$Zn_4O(BDC)_3$) where Fe(III) is partially substituted to the MOF. The synergic effect between Co_3O_4 and $ZnFe_2O_4$ in the nanocomposite material showed high specific capacitance and high energy density. Another rGO-wrapped MoO_3 composite was prepared by annealing a mixture of Mo-based MOF and GO sheets in different atmosphere. The composite showed higher performance than other solid-state flexible supercapacitor devices.

3.3.4 MOF-DERIVED CARBON MATERIALS FOR SUPERCAPACITORS

MOF-derived carbon materials have large specific surface area and pore volume which allows fast ion diffusion thereby enhancing the interaction between the electrolyte and electrode [20]. Therefore, these materials have opportunities to be used in supercapacitors as electrodes.

By carbonization of MOF, carbon materials with remarkable electrochemical performance and cycling stability can be produced. Further, they can be activated by KOH to increase the specific surface area and pore volume. Moreover, materials derived from MOF doped with nitrogen, increase their wettability, electrical conductivity, and the specific capacitance by a pseudocapacitive effect. One such material is hierarchical nitrogen-doped porous carbon (HNPC) derived by carbonization from isoreticular MOF (IRMOF-3) containing nitrogen. The specific area and the nitrogen content are temperature dependent. The pseudocapacitance is improved by nitrogen doping as the specific surface area and pore volume of HNPC is enhanced at 950C [20] and hence had high capacitance. But the stability of the material was very low and only amorphous carbon was obtained. Another example is ZIF-8@ZIF-67 crystals, a core-shell structured MOF from which graphitic carbon and nitrogen-doped carbon was obtained. The composite showed distinguished high capacitance and is used for energy storage. These electrochemical double-layer capacitors discussed are applied in different electronic devices and other applications.

TABLE 3.5 Summary of Metal-Organic Framework (MOF)-Based Electrodes, Their Energy Density, Power Density, Specific Capacitance and Current Densities in Supercapacitors

Electrode Material	Current Density (A/g)	Specific Capacity (F/g)	Energy Density (Wh kg^{-1})	Power Density (W kg^{-1})	Ref
NPCS (ZIF-8@ZnO)	1	290	6.12	250	[85]
Ni/Co-MOF-rGO	1	860	72.8	42.5 (K)	[20]
Ni-MOF/CNT	0.5	1765	36.6	480	[20]
RuO_2/PCs (ZIF-8)	1	539.6	23.38	600	[86]
Zn-doped Ni-MOF	0.25	1620	27.56	1750	[20]
Co-MOF	1	179.2	31.4	5640	[57]
Ce-MOF/GO	1	2221.2	11.96	4497	[87]
SGO/ZIF-67/POAP	1	825	412.5	3140.86	[88]
ZIF-67/POAP	1	725	362	3102.85	[89]

3.4 METAL-ORGANIC FRAMEWORKS IN DYE-SENSITIZED SOLAR CELLS

Solar cells are photoelectric conversion devices that use sunlight to produce electricity. Semiconductor-based dye-sensitized solar cells (DSSCs) are becoming popular due to their low cost and high efficiency in charge separation and have the potential to bring about effective power conversion, thereby solving global energy crisis. The semiconducting property of MOFs can be customized by tuning the size of metal nodes, exchanging the organic ligands, and altering the coordination modes between the ligand and the metal. Since MOF-based materials have the outstanding ability to harvest light and possess large surface area, they can be used to sensitize the electrode by serving as photoanodes/auxiliaries in DSSCs.

The photovoltaic cell fabricated using $Al_2(BDC)_3$ exhibited a very good performance than the individual components. This was attributed to the organic guests introduced into $Al_2(BDC)_3$ act as trapping moiety for electrons or holes for longer lifetime of charge separation.

Another hybrid material made up of ZIF-8-coated TiO_2 with varied thicknesses has been synthesized by mixing TiO_2, $Zn(NO_3)_2$, and 2-methyl imidazole in methanol and was applied for DSSC. By varying the reaction time, the thickness of the hybrid material can be controlled. The adsorption of dyes on the ZIF-8-coated layer increased with increasing thickness of the material. The open circuit voltage was also greater for the thickly coated ZIF-8. The enhanced dye adsorption, faster electron transport, and better charge collection efficiency in the as-synthesized ZIF-8-TiO_2 hybrid are mainly responsible for the enhanced performance. MOF-derived materials are also successfully used as photoanode in DSSC with improved efficiency. Materials such as MIL-125-TiO_2 composite, Co-MOF/TiO_2 composite, Zr-based MOF-525 in the MOF/perovskite composite increased the power conversion efficiency when used in the construction of solar cells. Redox centers were introduced into the MOF-based solar cells as sensitizing materials. For example, RuDCBPY (RuDCBPY = ruthenium(II)bis-(2,2′-bipyridine)(2,2′-bipyridine-5,5′-dicarboxylic acid)) containing UiO-67 type MOFs were grown as thin films on TiO_2 as sensitizing materials.

3.5 MOF-BASED MATERIALS FOR FUEL CELLS

A fuel cell is an electrochemical transformation device that has a consistent inventory of fuel, like hydrogen, petroleum gas, or methanol, and oxidant like oxygen, air, or hydrogen peroxide. Each power device comprises of an anode, a cathode, and electrolyte which permit ions to move between the two sides of the energy fuel cells. Polymer electrolyte membrane fuel cells (PEMFCs) due to their current applications have gained more attraction (Figure 3.14). Nafion is one of the popular PEM material, which possess high proton conductivity and thermal and chemical stability but due to their temperature dependence performance and requires humid environment which limits its efficiency, a new electrolyte material with high proton conductivity is required. The high proton conductivities of emerging MOFs at low- and high-temperature with low cost synthesis make them promising electrolytes in fuel cells.

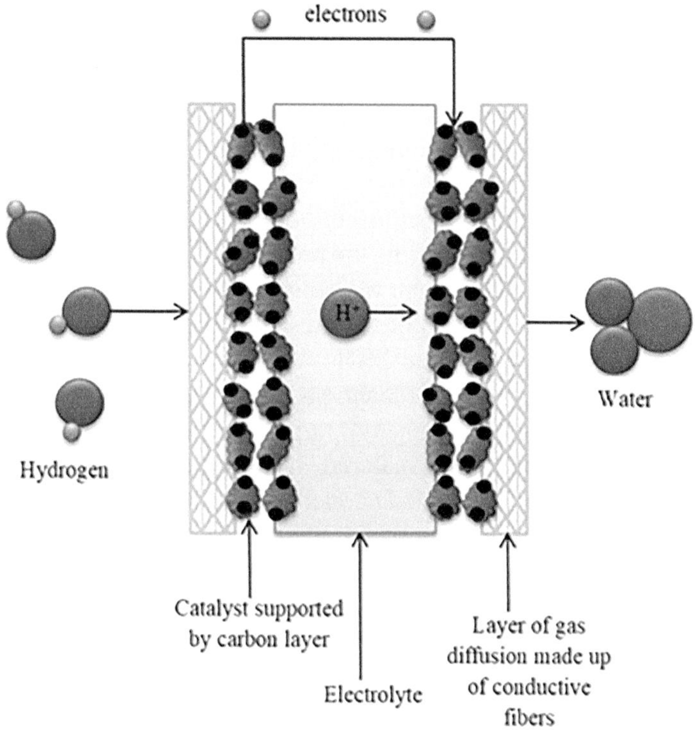

FIGURE 3.14 The configuration of a single polymer electrolyte membrane (PEM) cell.

3.5.1 MOFS FOR PROTON CONDUCTION AT LOW TEMPERATURE

$\{(NH_4)_2(adp)[Zn_2(ox)_3]\cdot 3H_2O\}_n$ (ox = oxalic acid, adp = adipic acid) is the first proton-conducting MOF with a comparable performance to that of Nafion at low temperature, where carboxylic group in adipic acid and ammonium ions serve as conducting media. Under relative humidity, the proton conduction was high for this MOF. The proton conduction is facilitated by the presence of functional groups in the linkers, water, and guest molecules coordinated to the metal ion in the MOF structure. The use of 2D porous frameworks with oxalate moiety as a host led to the investigation of high proton conductors by introducing different carriers of proton into the pores of MOF. Introduction of different substituents not only controls the pore size of the MOFs but also allows effective proton conduction by adjusting the species, concentration, and mobility of proton carriers. Ligands like sulfonic acids and organic phosphonic acids provide facile pathway for proton conduction by bridging with the solvent water molecules through hydrogen bonding.

The highly ordered and defects in the arrangement of MOF also provide significant change in proton conduction. For example, in a zirconium-sulfoterephthalate MOF, the effect of defect on proton conduction was investigated. The defect in the MOF was introduced by adding an excess amount of (1) acetic acid, (2) sulfoacetic acid, and (3) soaking the normal MOF in 0.1 M sulfuric acid during the preparation process, and the proton conductivity was compared with the ordered MOF. It was found that the samples to which defects was introduced showed high proton conductivity than the corresponding ordered MOF. Another example to illustrate the high proton conductivity of ordered MOF is $\{Cu(TCPP)\}_n$ (H_2TCPP = 5,10,15,20-tetrakis(4-carboxyphenyl)porphyrin) nanofilm. The abnormally high conductivity of this material is due to the presence of many free groups, like coordinated water and noncoordinated carboxyl group, on the surface of the nanosheet providing highly oriented crystalline structure to the MOF thin film. Although crystalline or film forms of MOFs conduct protons with very high conductivity at low temperature, their assembly in fuel cells is not much explored.

3.5.2 *MOFS FOR PROTON CONDUCTION AT HIGH TEMPERATURE*

MOFs are successfully developed to conduct protons at temperature greater than 100°C embedding guest molecules like H_2SO_4, H_3PO_4, and nonvolatile organic molecules such as triazole, imidazole, benzimidazole, and histamine. They set up easy proton delocalization pathway to promote proton conduction in anhydrous condition. Few examples are aluminum-based MOFs: Im@$\{Al(\mu_2\text{-}OH)(1,4\text{-}NDC)\}_n$ (1,4-NDC = 1,4-naphthalenedicarboxylate) and Im@$\{Al(\mu_2\text{-}OH)(1,4\text{-}BDC)\}_n$ [90], sulfonate-based MOF: β-PCMOF2$(Tz)_x$. The dynamic movement of protons is enhanced by the presence of imidazole (Im) molecule in Al MOF and triazole (Tz) molecule in the sulfato MOF that encapsulated into the pores. Among the series of β-PCMOF2$(Tz)_x$, where x = 0.30, 0.45, and 0.60, β-PCMOF2$(Tz)_{0.45}$ showed the highest proton conductivity and is integrated into a membrane-electrode assembly to fabricate H_2-air cell. The emf of this cell recorded was found to be stable which showed the effectiveness of proton conduction by guest molecules incorporated into MOFs structure under anhydrous condition and at higher temperature.

This paved ways to explore new materials to operate as MOF-based fuel cells at temperature above 100°C. Similarly a dry H_2-air cell with membrane-electrode assembly has been manufactured in which phosphoric acid molecules are used as starting material to form a Zn-based MOF $\{Zn(H_2PO_4)_2(HTz)_2\}_n$ (HTz = 1,2,4-triazole. The construction of this fuel cell applies a low-cost strategy for the synthesis of new proton conductors. There are rare materials available which display high conductivity in both low- and high-temperature regions. Among them oxalate based MOF behave as excellent proton conductors. One such example is $\{[(Me_2NH_2)_3(SO_4)]_2[Zn_2(ox)_3]\}_n$ a Zn-based MOF.

3.5.3 OXYGEN REDUCTION REACTION

The working of PEMFC is as follows:

1. H_2 which acts as a fuel is delivered to the anode. Oxidation takes place to generate protons and electrons at the anode [47].
2. The protons are passed through the electrolyte and reach the cathode. While the electrons move to the external circuit for supplying current.
3. At the cathode, O_2 combines with protons and electrons in the presence of a catalyst to produce water, and heat is released.

The advantages of using PEMFCs are:

1. Excellent energy density.
2. Fast refueling.
3. Clean point of operation emissions.

Due to their remarkable properties, they have a variety of applications. The application of H_2-fuelled PEMFCs in automobiles would drastically cut short the emission of greenhouse gas (CO_2) produced every year. But due to their high cost and insufficient durability, they are limited in the automotive market.

Another drawback encountered is the use of Pt catalyst for ORR which is very important in a fuel cell as well as for metal-air batteries. In PEMFCs, Pt is the most efficient electrocatalyst due to its corrosion resistance and exceptional catalytic and electrical properties [47]. But the expense of Pt catalyst is 50% the expense of membrane electrode assembly. Therefore, it is highly necessary to develop efficient and cost-effective materials to replace Pt-based catalysts for PEMFCs. By suitably selecting clusters of nonprecious, electrochemically active metals such as Co, Fe, Cu, and multifunctional linkers, the structure of MOFs can be tuned and applied as an electrocatalyst in PEMFCs.

The efficiency of MOFs and MOF-derived nanostructured materials as electrocatalysts has been reported in recent years. The first example of MOF as an electrocatalyst for ORR is [copper(II) benzene-1,3,5-tricarboxylate] and [copper(II)-2,2′-bipyridinebenzene-1,3,5-tricarboxylate] [91], two Cu-based MOFs. The electrocatalytic activity of [copper(II)-2,2′-bipyridinebenzene-1,3,5-tricarboxylate] MOF toward four-electron reduction of O_2 was excellent and stable. Generally, Cu-MOFs showed excellent catalytic performance but limited electron conduction. Hence, GO-incorporated copper MOF composites were designed to improve stability and catalytic activity. The GO Cu-MOFs prepared by Maryam and co-workers [91] exhibited excellent tri-catalytic activity for ORR, OER, and HER and also showed remarkable stability in acidic media when compared with pure Cu-MOF. Also, maximum power density was achieved for the GO–Cu–MOF hybrid. Therefore, GO-Cu-MOFs which is easily synthesized at low cost and showed excellent electrocatalytic behavior hopefully can replace Pt catalysts in fuel cells. Similarly, Fe-MOF with reduced GO has been employed as ORR catalyst. The synergistic effect of rGO with Fe-porphyrin catalysts affords four electrons ORR [92].

3.6 COST EFFECTIVENESS OF MOFS

A method of synthesizing safe MOFs considering the economic feasibility without relying on expensive or rare raw materials has been recently studied. It is known that MOF has advantages in cost-effectiveness, environmental friendliness, easy access, safety, and stability compared to other existing energy storage materials [93–96]. The direct cause of expenses in the manufacture of MOFs is linker costs, space time yield, down-stream processing effort, and solvents used. The BASF team has described the following as evaluation criteria for commercialization of MOFs [97].

1. Large-scale production
2. Toxicity of reagents and solvents
3. Cost of raw materials
4. Availability and purity of chemicals
5. Safety

By meeting these criteria, MOFs can be designed with cost-effective and environmentally friendly compared to other energy storage materials, and they also can be advantageous in terms of stability.

3.7 CHALLENGES, PROSPECTS, AND OPPORTUNITIES

MOFs and MOF-derived materials with controllable morphology, flexible structure, large pore size, large surface area, and active metal sites [20] can be used in batteries, supercapacitors, solar cells, and fuel cells as potential electrode materials for energy storage applications [20]. The cyclability, performance, and reaction kinetics of MOFs and their derivatives can be improved toward energy storage and conversion by modifying the structural features [1, 2]. Yet, the real-time utility of MOFs and MOF-derived materials is still a challenge, because of their low stability toward moist air, poor electrical conductivity, catalytic performance, interaction between MOFs and polysulfides/polyselenides, etc. The incorporation of carbon-based materials into MOFs and MOF-derived materials is a hopeful approach to meet out these drawbacks. It is important to understand the mechanism of MOFs and MOF-derived materials for their applicability in energy storage [98]. There are also problems which arise during the scaling up of the MOF production. In a scale-up approach, the organic solvents used for the synthesis may present corrosivity and toxicity. They may also suffer from poor recyclability and may be flammable. Sometimes the accumulation of counter ions like nitrates may explode, and corrosive by-products may be generated from chloride salts. Hence metal oxides and metal hydroxides are highly preferred in bulk production. Nonvolatile solvents and unreacted raw materials in should be removed from the MOF pores before they are used for any application. The organic ligands required for the synthesis of MOFs should not be complicated and should be customized. It is very important to maintain the size of the particles based on their applications. For example, nano-sized particles are required for thin film applications, while larger particles are needed for storage applications to preserve their stationary state [99]. Now many novel functional MOFs have been commercialized and are in progress for practical implementation. So MOFs can serve as an essential material, competing with today's well-known industrial materials, thus qualifying the global pressure on energy and the environment. Day-to-day production and utilization of materials applied in daily life will create high interest and more research in the development of MOF-based materials.

3.8 CONCLUSION

Batteries with high energy density and supercapacitors with high power density have pulled in much consideration since they are essential to the power supply of future versatile electronic devices, electric cars, automated ethereal vehicles, and so on. The vital components for batteries and supercapacitors are the electrode materials, which influence the practical energy and power density. MOFs owning novel morphology, high explicit surface zone, well-designed linkers, and metal sites are incredible electrode materials for electrochemical energy storage devices. In this chapter, we have consolidated the recent progress in MOF-based lithium-based batteries, sodium-ion batteries, supercapacitors, dye sensitized solar cells and fuel cells. As research interest for using MOFs in energy applications increases, further sustained and intense efforts in this field will increase the outlook of MOFs, composites based on MOFs, and derivatives of MOF for energy applications.

KEYWORDS

- **renewable energy**
- **electrochemical energy storage**
- **functional materials**
- **metal-organic frameworks**

REFERENCES

1. Huggins, R.A. *Energy Storage Fundamentals—Materials and Applications*, 2nd edn., Springer International Publishing: Switzerland, 2016, p. 119. doi: 10.1007/978-3-319-21239-5.
2. Baumann, A.E., Burns, D.A., Liu, B., Thoi, S.V. Metal-organic framework functionalization and design strategies for advanced electrochemical energy storage devices. *Communications Chemistry* 2019, 2, 86. doi: 10.1038/s42004-019-0184-6.
3. Wang, L., Han, Y., Feng, X., Zhou, J., Qi, P., Wang, B. Metal-organic frameworks for energy storage: Batteries and supercapacitors. *Coordination Chemistry Reviews* 2016, 307, 361–381. doi: 10.1016/j.ccr.2015.09.002.

4. Moseley, P.T., Garche, J. *Electrochemical Energy Storage for Renewable Sources and Grid Balancing*, Elsevier: Netherlands, 2015, p. 381.
5]5. Lv, W., Li, Z., Deng, Y., Yang, Q.H., Kang, F. Graphene-based materials for electrochemical energy storage devices: Opportunities and challenges. *Energy Storage Materials* 2016, 2, 107–138. doi: 10.1016/j.ensm.2015.10.002.
6. Kaneti, Y.V., Zhang, J., He, Y.B., Wang, Z., Tanaka, S., Hossain, M.SA., Pan, Z.Z, Xiang, B., Yang, Q.H., Yamauchi, Y. Fabrication of an MOF-derived heteroatom-doped Co/CoO/carbon hybrid with superior sodium storage performance for sodium-ion batteries. *Journal of Materials Chemistry A* 2017, 5, 15356–15366. doi: 10.1039/c7ta03939e.
7. Wei, Q., Xiong, F., Tan, S., Huang, L., Lan, E.H., Dunn, B., Mai, L. Porous one-dimensional nanomaterials: Design, fabrication and applications in electrochemical energy storage. *Advanced Materials*, 2017, 29, 1602300. doi: 10.1002/adma.201602300.
8. Liu, Z., Yuan, X., Zhang, S., Wang, J., Huang, Q., Yu, N., Zhu, Y., Fu, L., Wang, F., Chen, Y., Wu, Y. Three-dimensional ordered porous electrode materials for electrochemical energy storage. *NPG Asia Materials* 2019, 11, 12. doi: 10.1038/s41427-019-0112-3.
9. Zu, L., Zhang, W., Qu, L., Liu, L., Li, W., Yu, A., Zhao, D. Mesoporous materials for electrochemical energy storage and conversion. *Advanced Energy Materials* 2020, 10, 2002152. doi: 10.1002/aenm.2002152.
10. Zhang, X., Cheng, X., Zhang, Q. Nanostructured energy materials for electrochemical energy conversion and storage: A review. *Journal of Energy Chemistry* 2016, 25, 967–984. doi: 10.1016/j.jechem.2016.11.003.
11. Gao, X., Dong, Y., Li, S., Zhou, J., Wang, L., Wang, B. MOFs and COFs for batteries and supercapacitors. *Electrochemical Energy Reviews* 2020, 3, 81–126. doi: 10.1007/s41918-019-00055-1.
12. Barsukov, I.V., Johnson, C.S., Doninger, J.E., Barsukov, V.Z. *New Carbon Based Materials for Electrochemical and Fuel Cells Energy Storage Systems: Batteries, Supercapacitors, NATO Science Series*, Springer: The Netherlands, Vol, 229, 2006.
13. Siwal, S.S., Zhang, Q., Devi, N., Thakur, V.K. Carbon-based polymer nanocomposite for high-performance energy storage applications. *Polymers* 2020, 12, 505. doi: 10.3390/polym12030505.
14. Kharbachi, A.E., Dematteis, E.M., Shinzato, K., Stevenson, S.C., Bannenberg, L.J., Heere, M., Zlotea, C., Szilágyi, P.Á., Bonnet, J.P., Grochala, W., Gregory, D.H., Ichikawa, T., Baricco, M., Hauback, B.C., Metal hydrides and related materials. Energy carriers for novel hydrogen and electrochemical storage. *Journal of Physical Chemistry C* 2020, 124, 7599–7607. doi: 10.1021/acs.jpcc.0c01806.
15. Fang, S., Bresser, D., Passerini, S. Transition metal oxide anodes for electrochemical energy storage in lithium- and sodium-ion batteries. *Advanced Energy Materials* 2019, 10, 1902485. doi: 10.1002/aenm.201902485.
16. Gao, B., Li, X., Ding, K., Huang, C., Li, Q., Chuc, P.K., Huo, K. Recent progress in nanostructured transition metal nitrides for advanced electrochemical energy storage. *Journal of Materials Chemistry A* 2019, 7, 14–37. doi: 10.1039/c8ta05760e.
17. Kempahanumakkagari, S., Vellingiri, K., Deep, A., Kwon, E.E., Bolan, N., Kim, K.H. Metal-organic framework composites as electrocatalysts for electrochemical sensing

applications. *Coordination Chemistry Reviews* 2018, 357, 105–129. doi:10.1016/j.ccr.2017.11.028.
18. Remya, V.R., Kurian, M. Synthesis and catalytic applications of metal-organic frameworks: A review on recent literature. *International Nano Letters* 2019, 9, 17–29. doi: 10.1007/s40089-018-0255-1.
19. Shrivastav, V., Sundriyal, S., Goel, P., Kaur, H., Tuteja, S.K., Vikrant, K., Kim, K.H., Tiwari, U.K., Deep, A. Metal-organic frameworks (MOFs) and their composites as electrodes for lithium battery applications: Novel means for alternative energy storage. *Coordination Chemistry Reviews* 2019, 393, 48–78. doi: 10.1016/j.ccr.2019.05.006.
20. Xu, G., Nie, P., Dou, H., Ding, B., Li, L., Zhang, X. Exploring metal organic frameworks for energy storage in batteries and supercapacitors. *Materials Today* 2017, 20, 191–209. doi: 10.1016/j.mattod.2016.10.003.
21. Sang, A.H., Hamzeh, Q., Jong-Won, L., Min-Sik, P., Jung, H.K. Metal-organic framework derived porous structures towards lithium rechargeable batteries. *EcoMat* 2023, 5, e12283. doi: 10.1002/eom2.12283.
22. Seoane, B., Castellanos, S., Dikhtiarenko, A., Kapteijn, F., Gascon, J. Multi-scale crystal engineering of metal organic frameworks. *Coordination Chemical Review* 2016, 307, 147–187.
23. Bon, V., Kavoosi, N., Senkovska, I. Tuning the flexibility in MOFs by SBU functionalization. *Dalton Transactions* 2016, 45(10), 4407–4415.
24. Halder, A., Ghoshal, D. Structure and properties of dynamic metal-organic frameworks: A brief accounts of crystalline-to-crystalline and crystalline-to-amorphous transformations. *CrstEngComm* 2018, 20(10), 1322–1345.
25. Sun, L., Hendon, C.H., Minier, M.A., Walsh, A., Dinca, M. Million-fold electrical conductivity enhancement in Fe_2(DEBDC) versus Mn_2(DEBDC) (E = S, O). *Journal of American Chemical Society* 2015, 137(19), 6164–6167.
26. Kung, C.-W., Platero-Prats, A.E., Drout, R.J. Inorganic "conductive glass" approach to rendering mesoporous metal-organic frameworks electronically conductive and chemically responsive. *ACS Applied Material Interfaces* 2018, 10(36), 30532–30540.
27. Wang, T.C., Hod, I., Audu, C.O. Rendering high surface area, mesoporous metal-organic frameworks electronically conductive. *ACS Applied Material Interfaces* 2017, 9(14), 12584–12591.
28. Phang, W.J., Jo, H., Lee, W.R. Superprotonic conductivity of a UiO-66 framework functionalized with sulfonic acid groups by facile postsynthetic oxidation. *Angewandte Chemie International Edition* 2015, 54(17), 5142–5146.
29. Kim, S.R., Joarder, B., Hurd, J.A. Achieving superprotonic conduction in metal-organic frameworks through iterative design advances. *Journal of American Chemical Society* 2018, 140(3), 1077–1082.
30. Sun, L., Hendon, C.H., Park, S.S. Is iron unique in promoting electrical conductivity in MOFs? *Chemical Science* 2017, 8(6), 4450–4457.
31. Shen, L., Wu, H.B., Liu, F. Creating lithium-ion electrolytes with biomimetic ionic channels in metal-organic frameworks. *Advanced Material* 2018, 30(23), 1707476.
32. Eddaoudi, M., Kim, J., Rosi, N. Systematic design of pore size and functionality in isoreticular MOFs and their application in methane storage. *Science* 2002, 295(5554), 469–472,

33. Chen, Z., Hanna, S.L., Redfern, L.R., Alezi, D., Islamoglu, T., Farha, O.K. Reticular chemistry in the rational synthesis of functional zirconium cluster-based MOFs. *Coordination Chemical Reviews* 2019, 386, 32–49.
34. Rubio-Martinez, M., Avci-Camur, C., Thornton, A.W., Imaz, I., Maspoch, D., Hill, M.R. New synthetic routes towards MOF production at scale. *Chemical Society Reviews* 2017, 46(11), 3453–3480.
35. Silva, P., Vilela, S.M.F., Tomé, J.P.C., Paz, FAA. Multifunctional metal-organic frameworks: From academia to industrial applications. *Chemical Society Reviews* 2015, 44(19), 6774–67803.
36. Ren, J., Dyosiba, X., Musyoka, N.M., Langmi, H.W., Mathe, M., Liao, S. Review on the current practices and efforts towards pilot-scale production of metal-organic frameworks (MOFs), *Coordination Chemical Reviews* 2017, 352, 187–219.
37. Isaeva, V.I., Kustov, L.M. Microwave activation as an alternative production of metalorganic frameworks. *Russian Chemical Bulletin* 2013, 65(9), 2103–2114.
38. Shekhah, O., Wang, H., Zacher, D., Fischer, R.A., Wöll, C. Growth mechanism of metal-organic frameworks: Insights into the nucleation by employing a step-by-step route. *Angewandte Chemie International Edition* 2009, 48(27), 5038–5041.
39. Meilikhov, M., Yusenko, K., Schollmeyer, E., Mayer, C., Buschmann, H.-J., Fischer, R.A. Stepwise deposition of metal organic frameworks on flexible synthetic polymer surfaces. *Dalton Transactions* 2011, 40(18), 4838–4841.
40. Shekhah, O., Liu, J., Fischer, R.A., Wöll, C. MOF thin films: Existing and future applications. *Chemical Society Reviews* 2011, 40(2), 1081–1106.
41. Zhu, J.P., Wang, X.H., Zuo, X.X. The application of metal-organic frameworks in electrode materials for lithium–ion and lithium–sulfur batteries. *Royal Society Open Science* 2019, 6, 190634. doi: 10.1098/rsos.190634.
42. Saravanan, K. Nagarathinam, M., Balaya, P., Vittal, J.J. Lithium storage in a metal organic framework with diamondoid topology—a case study on metal formats. *Journal of Materials Chemistry* 2010, 20, 8329–8335. doi:10.1039/c0jm01671c
43. Ke, F.-S., Wu, Y.-S., Deng, H. Metal-organic frameworks for lithium ion batteries and supercapacitors. *Journal of Solid State Chemistry* 2015, 223, 109–121. doi: 10.1016/j.jssc.2014.07.008
44. Zhu, J.P., Chen, X., Wang, X., Zuo, X.X., Li, J. Disc-shaped $Li_{4-x}K_xTi_5O_{12}$ derived from MIL-125(Ti) as an anode material with high performance for lithium-ion batteries. *Journal of Electronic Materials* 2021, 50(7), 4066–4074.
45. Zhao, Y., Song, Z., Li, X., Sun, Q., Cheng, N., Lawes, S., Sun, X. Metal organic frameworks for energy storage and conversion. *Energy Storage Materials* 2016, 2, 35–62. doi: 10.1016/j.ensm.2015.11.005.
46. Shaoju, J., Xiaofan, M., Qingming, W., Jinglin, W., Yifan, W., Shaohua, J., Wenhui, X., Weisen, Y. Hierarchical porous Co_3O_4 nanocages with elaborate microstructures derived from ZIF-67 toward lithium storage. *Vacuum* 2021, 184, 109879. doi: 10.1016/j.vacuum.2020.109879.
47. Zhang, L., Wu, H.B., Madhavi, S., Hng, H.H., Lou, X.W. Formation of Fe_2O_3 microboxes with hierarchical shell structures from metal-organic frameworks and their lithium storage properties. *Journal of the American Chemical Society* 2012, 134, 17388–17391. doi: 10.1021/ja307475c.

48. Jia, D., Tong, R., Ning, L. BN nanosheets in situ mosaic on MOF-5 derived porous carbon skeleton for high-performance lithium-ion batteries. *Journal of Alloys and Compounds* 2021, 857, 157571.
49. Ma, J., Kong, Y., Liu, S., Li, Y., Jiang, J., Zhou, Q., Huang, Y., Han, S. Flexible phosphorus-doped graphene/metal-organic framework-derived porous Fe_2O_3 anode for lithium-ion battery. *ACS Applied Energy Materials* 2020, 3(12). doi:10.1021/acsaem.0c02030.
50. Eskandari, M., Shahbazi, N., Marcos, A.V., Malekfar, R., Taboada, P. Facile MOF-derived $NiCo_2O_4$/r-GO nanocomposites for electrochemical energy storage applications. *Journal of Molecular Liquids* 2022, 348, 118428.
51. Peng, Z., Yi, X., Liu, Z., Shang, J., Wang, D. Triphenylamine-based metal-organic frameworks as cathode materials in lithium ion batteries with coexistence of redox active sites, high working voltage, and high rate stability. *ACS Applied Material Interfaces* 2016, 8(23), 14578–14585.
52. Wang, X.-J., Krumeich, F., Nesper, R. Nanocomposite of manganese ferrocyanide and graphene: A promising cathode material for rechargeable lithium ion batteries. *Electrochemistry Communications* 2013, 34, 246–249.
53. Hu, X., Lou, X., Li, C. Facile synthesis of the Basolite F300-like nanoscale Fe-BTC framework and its lithium storage properties. *RSC Advances* 2016, 6(115), 114483–114490.
54. Li, C., Lou, X., Shen, M. High anodic performance of Co 1,3,5-benzenetricarboxylate coordination polymers for Li-ion battery. *ACS Applied Material Interfaces* 2016, 8(24), 15352–15360.
55. Xiong, P., Zeng, G., Zeng, L., Wei, M. Prussian blue analogs Mn $[Fe(CN)_6]_{0.6667}nH_2O$ cubes as an anode material for lithium ion batteries. *Dalton Transactions* 2015, 44(38), 16746–16751.
56. Hu, H., Lou, X., Li, C. A thermally activated manganese 1,4-benzenedicarboxylate metal organic framework with high anodic capability for Li-ion batteries. *New Journal of Chemistry* 2016, 40(11), 9746–9752.
57. Fang, G., Zhou, J., Liang, C. MOFs nanosheets derived porous metal oxide-coated three-dimensional substrates for lithium-ion battery applications. *Nano Energy* 2016, 26, 57–65.
58. Zhang, X., Li, D., Zhu, G., Lu, T., Pan, L. Porous $CoFe_2O_4$ nanocubes derived from metal-organic frameworks as high performance anode for sodium ion batteries. *Journal of Colloid Interface Science* 2017, 499, 145–150.
59. Pham, H.T.T., Yun, J., Kim, S.Y. Nanoarchitectonics of the cathode to improve the reversibility of Li-O_2 batteries. *Beilstein Journal of Nanotechnology* 2022, 13, 689–698.
60. Sui, Z.-Y., Zhang, P.-Y., Xu, M.-Y., Liu, Y.-W., Wei, Z.-X., Han, B-H. Metal-organic framework-derived metal oxide embedded in nitrogen-doped graphene network for high-performance lithium-ion batteries. *ACS Applied Material Interfaces* 2017, 9(49), 43171–43178.
61. Du, M., Rui, K., Chang, Y. Carbon necklace incorporated electroactive reservoir constructing flexible papers for advanced lithium-ion batteries. *Small* 2018, 14(2), 1702770.

62. Xie, X.C., Huang, K.J., Wu, X. Metal-organic framework derived hollow materials for electrochemical energy storage. *Journal of Materials Chemistry A* 2018, 6, 6754–6771. doi: 10.1039/c8ta00612a.
63. Yin, W., Shen, Y., Zou, F., Hu, X., Chi, B., Huang, Y. Metal-organic framework derived ZnO/ZnFe2O4/C nanocages as stable cathode material for reversible lithium–oxygen batteries. *ACS Applied Material Interfaces* 2015, 7(8), 4947–4954.
64. Chen, W., Zhang, Z., Bao, W. Hierarchical mesoporous γ-Fe_2O_3/carbon nanocomposites derived from metal organic frameworks as a cathode electrocatalyst for rechargeable LiO_2 batteries. *Electrochimica Acta* 2014, 134, 293–301.
65. Gan, Y., Lai, Y., Zhang, Z., Chen, W., Du, K., Li, J. Hierarchical Cr_2O_3@OPC composites with octahedral shape for rechargeable non-aqueous lithium-oxygen batteries. *Journal of Alloys and Compounds* 2016, 665, 365–372.
66. Yan, W., Guo, Z., Xu, H., Lou, Y., Chen, J., Li, Q. Downsizing metal-organic frameworks with distinct morphologies as cathode materials for high-capacity Li–O_2 batteries. *Materials Chemistry Frontiers* 2017, 1(7), 1324–1330.
67. Shin, S., Yoon, H., Yoon, Y., Park, S., Shin, M.W. Porosity tailoring of the Zn-MOF-5 derived carbon materials and its effects on the performance as a cathode for lithium-air batteries. *Microporous Mesoporous Materials* 2021, 311, 110726.
68. Manthiram, A., Fu, Y., Su, Y.S. Challenges and prospects of lithium-sulfur batteries. *Accounts of Chemical Research* 2013, 46, 1125–1134. doi: 10.1021/ar300179v.
69. Zhao, R., Liang, Z., Zou, R., Xu, Q. Metal-organic frameworks for batteries. *Joule* 2018, 2, 2235–2259. doi: 10.1016/j.joule.2018.09.019.
70. Qi, C., Xu, L., Wang, J. Titanium-containing metal-organic framework modified separator for advanced lithium–sulfur batteries. *ACS Sustainable Chemical Engineering* 2020, 8(34), 12968–12975.
71. Jin, H.-G., Wang, M., Wen, J-X. Oxygen vacancy-rich mixed-valence cerium MOF: An efficient separator coating to 76 of 80 high-performance lithium–sulfur batteries. *ACS Applied Material Interfaces* 2021, 13(3), 3899–3910.
72. Wang, Q., Zhao, H., Li, B. MOF-derived Co_9S_8 nano-flower cluster array modified separator towards superior lithium sulfur battery. *Chinese Chemical Letters* 2021, 32(3), 1157–1160.
73. Jiang, H., Liu, X.-C., Wu, Y. Metal-organic frameworks for high charge–discharge rates in lithium–sulfur batteries. *Angewandte Chemie International Edition* 2018, 57(15), 3916–3921.
74. Su, D., Cortie, M., Fan, H., Wang, G. Prussian blue nanocubes with an open framework structure coated with PEDOT as high-capacity cathodes for lithium–sulfur batteries. *Advanced Materials* 2017, 29(48), 1700587.
75. Li, Z., Li, C., Ge, X. Reduced graphene oxide wrapped MOFs-derived cobalt-doped porous carbon polyhedrons as sulfur immobilizers as cathodes for high performance lithium sulfur batteries. *Nano Energy* 2016, 23, 15–26.
76. Baumann, A.E., Han, X., Butala, M.M., Thoi, V.S. Lithium thiophosphate functionalized zirconium MOFs for Li-S batteries with enhanced rate capabilities. *Journal of American Chemical Society* 2019, 141(44), 17891–17899.
77. Bai, S., Liu, X., Zhu, L., Wu, S., Zhou, H. Metal-organic framework-based separator for lithium–sulfur batteries. *Nature Energy* 2016, 1, 16094.

78. Chen, H., Xiao, Y., Chen, C. Conductive MOF-modified separator for mitigating the shuttle effect of lithium–sulfur battery through a filtration method. *ACS Applied Material Interfaces* 2019, 11(12), 11459–11465.
79. Kim, S.H., Yeon, J.S., Kim, R., Choi, K.M., Park, H.S. A functional separator coated with sulfonated metal-organic framework/Nafion hybrids for Li-S batteries. *Journal of Material Chemistry A* 2018, 6(48), 24971–24978.
80. Hong, X.-J., Song, C.-L., Yang, Y. Cerium based metal–organic frameworks as an efficient separator coating catalyzing the conversion of polysulfides for high performance lithium–sulfur batteries. *ACS Nano* 2019, 13(2), 1923–1931.
81. Mehta, T., Yasin, G., Arif, M., Shakeel, M., Korai, R.M., Nadeem, M., Muhammad, N., Lu, X. Metal-organic frameworks for energy storage devices: Batteries and supercapacitors. *Journal of Energy Storage* 2019, 21, 632–646. doi: 10.1016/j.est.2018.12.025.
82. Wang, M., Huang, Y., Zhu, Y. Synthesis of porous $Zn_xCo_{3-x}O_4$ hollow nanoboxes derived from metal-organic frameworks for lithium and sodium storage. *Electrochimica Acta* 2020, 335, 135694.
83. Hou, C.C., Xu, Q., Metal-organic frameworks for energy. *Advanced Energy Materials* 2019, 9, 1801307. doi: 10.1002/aenm.201801307.
84. Yujuan, C., Wu, Y., Mingyue, W., Ning, W., Longwen, Z., Qixia, G., Hao, G. Metal-organic frameworks as highly efficient electrodes for long cycling stability supercapacitors. *International Journal of Hydrogen Energy* 2021, 46, 18179–18206. doi: 10.1016/j.ijhydene.2021.03.003.
85. Huang, J., Hao, F., Zhang, X., Chen, J. N-doped porous carbon sheets derived from ZIF-8: Preparation and their electrochemical capacitive properties. *Journal of Electroanalytical Chemistry* 2018, 810, 86–94. doi: 10.1016/j.jelechem.2017.12.078.
86. Zhang, Q., Hu, Z., Yang, Y., Zhang, Z., Wang, X., Yang, X., An, Y., Guo, B. Metal organic frameworks-derived porous carbons/ruthenium oxide composite and its application in supercapacitor. *Journal of Alloys and Compounds* 2018, 735, 1673–1681. doi: 10.1016/j.jallcom.2017.11.268.
87. Ramachandran, R., Xuan, W., Zhao, C., Leng, X., Sun, D., Luo, D., Wang, F. Enhanced electrochemical properties of cerium metal-organic framework based composite electrodes for high-performance supercapacitor application. *RSC Advances* 2018, 8, 3462–3469. doi: 10.1039/C7RA12789H.
88. Boorboor, A.F., Kowsari, E., Ehsani, A. Ternary nanocomposites of conductive polymer/functionalized GO/MOFs: Synthesis, characterization and electrochemical performance as effective electrode materials in pseudocapacitors. *Journal of Solid State Chemistry* 2018, 265, 155–166.
89. Boorboor, A.F., Kowsari, E., Ehsani, A. P-type conductive polymer/zeolitic imidazolate framework-67 (ZIF-67) nanocomposite film: Synthesis, characterization, and electrochemical performance as efficient electrode materials in pseudocapacitors. *Journal of Colloid Interface Science* 2018, 509, 189–194.
90. Fu-Sheng, K., Yu-Shan, W., Hexiang, D. Metal-organic frameworks for lithium ion batteries and supercapacitors. *Journal of Solid State Chemistry* 2015, 223, 109–121. doi: 10.1016/j.jssc.2014.07.008

91. Maryam, J., Zhaolin, L., Kian, P. L. A graphene oxide and copper-centered metal organic framework composite as a tri-functional catalyst for HER, OER, and ORR. *Advanced Functional Material* 2013, 23(43), 5363–5372.
92. Wang, H., Zhu, Q.L., Zou, R., Xu, Q. Metal-organic frameworks for energy applications. *Chem* 2017, 2, 52–80.
93. Venturi, D.M., Campana, F., Marmottini, F., Costantino, F., Vaccaro, L. Extensive screening of green solvents for safe and sustainable UiO-66 synthesis. *ACS Sustainable Chemical Engineering* 2020, 8(46), 17154–17164.
94. Thomas-Hillman, I., Laybourn, A., Dodds, C., Kingman, S.W. Realising the environmental benefits of metal-organic frameworks: Recent advances in microwave synthesis. *Journal of Material Chemistry A* 2018, 6(25), 11564–11581.
95. Peh, S.B., Wang, Y., Zhao, D. Scalable and sustainable synthesis of advanced porous materials. *ACS Sustainable Chemical Engineering* 2019, 7(4), 3647–3670.
96. He, Q., Zhan, F., Wang, H., Xu, W., Wang, H., Chen, L. Recent progress of industrial preparation of metal-organic frameworks: Synthesis strategies and outlook. *Materials Today Sustainability* 2022, 17, 100104.
97. Teo, W.L., Zhou, W., Qian, C., Zhao, Y. Industrializing metalorganic frameworks: Scalable synthetic means and their transformation into functional materials. *Materials Today* 2021, 47, 170–186.
98. Wu, H.B., Lou, X.W. Metal-organic frameworks and their derived materials for electrochemical energy storage and conversion: Promises and challenges. *Science Advances* 2017, 3, eaap9252. doi: 10.1126/sciadv.aap9252.
99. An overview of metal-organic frameworks for green chemical engineering. *Engineering*, 2021, 7, 1115–1139. doi: 10.1016/j.eng.2021.07.001

CHAPTER 4

Surface Functionalized Nanofillers-Based Fluoropolymer Nanocomposites: Synthesis, Properties, and Applications

DIVYA THOMAS,[1*] DEEPU THOMAS,[2] EMMANUEL TOM,[2] and HA THANH TUNG[3]

[1]Department of Chemistry, St. Xavier's College Thumba, Kerala, India

[2]Department of Physics, Nirmalagiri College, Kerala, India

[3]Vinh Long University of Technology Education, Vietnam

*Corresponding author

ABSTRACT

Fluoropolymers are a type of polymer in which the hydrogen atom is replaced with fluorine. They have outstanding thermal, mechanical, electrical, and optical properties. The surface of a fluoropolymer is generally functionalized with various nanofillers to improve its potential properties. Its properties vary depending on the characteristic features of the nanofiller used. Therefore, surface modification or functionalization of nanofillers allows tuning the properties of fluoropolymer nanocomposites for specific applications. Surface-engineered fluoropolymers are widely used in medical devices, biotechnology, optoelectronics, and other fields. In general, polymer nanocomposites can be synthesized using one of three methods: *in situ* polymerization, solution, and melt blending. The method adopted for the synthesis of fluoropolymer nanocomposites is decided by considering factors such as the nature of the polymer matrix, the nanofiller used, and the specific properties required

Advances in Energy Materials: Materials Fabrication, Properties and Applications.
Lishin T, Maciej Jaroszewski, and Sabu Thomas (Eds.)

DOI: 10.1201/9781779643674-4

for the polymers. The purpose of this chapter is to discuss fluoropolymer nanocomposites' potential as energy sources. It discusses several kinds of functionalized nanofillers, their functionalization techniques, various methods for synthesizing fluoropolymer nanocomposites based on functionalized nanofillers, their properties, and possible uses. Finally, significant technological issues and prospective developments will be explored.

4.1 INTRODUCTION

Fluoropolymer can be defined as a fluorocarbon-based polymer consisting of multiple carbon-fluorine bonds. Polytetrafluoroethylene (PTFE), is the first member of this family [1, 2]. The versatile and unique properties of fluoropolymers originate from the replacement of hydrogen by fluorine in the polymer structure. The low polarizability and high electronegativity of fluorine molecules create a highly hydrophobic surface for fluoropolymers. They also possess high thermal stability, low refractive index, low friction resistance, low surface energy, and nonstick properties [3, 4]. Fluorine is the most electronegative element. The C–F bond is the most polar and strongest in organic chemistry.

The high chemical stability of fluoropolymers can be attributed to the presence of strong bond strength of the central C–C bonds that can reach values of around 470 kJ mol^{-1} as compared with 350–380 kJ mol^{-1} for those found in hydrocarbons. In addition to carbon and fluorine, certain fluoropolymers also contain oxygen and chlorine. Examples include perfluoroalkoxy polymer and polychlorotrifluoroethylene. Partially fluorinated fluoropolymers are another class of fluoropolymers that contain hydrogen along with carbon and fluorine. Examples are ethylene tetrafluoroethylene copolymer, ethylene chlorotrifluoroethylene copolymer, polyvinylidene fluoride (PVDF), etc.

4.2 NANOCOMPOSITES

Nanocomposites (PNCs) may be defined as a composite consisting of a bulk matrix and a nanomaterial with at least one dimension smaller than 100 nm. They exhibit better mechanical properties, thermal and dimensional stability, gas permeability, fire and chemical resistance, optical and electrical properties, etc., and lower loading levels. Based on the type of matrix, nanocomposites are classified into ceramic matrix nanocomposites (CMNC), metal matrix nanocomposites (MMNC), and polymer matrix nanocomposites (PMNC). CMNC consists of a ceramic dispersion medium such as an assembly of nitrides, oxides, silicides, and borides along with metal. In MMNC, metal acts as the matrix, and reinforcement is done by ceramic. PMNC may be defined as a composite consisting of a polymer and a nanomaterial with at least one dimension smaller than 100 nm. In PMNC there exists (1) a strong linkage between polymer chain and surface of nanoparticles and (2) weak interaction between polymer matrix and surface of nanoparticles.

The fascinating aspect of the inclusion of nanofillers in nanocomposites is that the very low amount of filler enhances barrier properties, membrane separation, partial discharge resistance, dielectric losses, polymer blend compatibility, thermal conductivity, and electrical conductivity [5, 6]. The exceptional behavior of nanofillers may be due to the relatively high surface-to-volume ratio. An epitome polymer nanocomposite consists of an optimum amount of homogeneously dispersed nanoparticles in a polymer matrix, and it should not undergo agglomeration [7]. The factors influencing the improved properties of polymers with nanofillers are properties of the polymer matrix, nature and type of nanofiller, concentration of polymer and filler, particle size, particle orientation, and particle–matrix interface adhesion.

Based on the size, nanofillers can be classified into one nanoscale dimension, 1D (nanoplatelet), two nanoscale dimensions, 2D (nanofiber), and three nanoscale dimensions 3D (nanoparticulate). The amount of nanofiller used in a polymer nanocomposite has a determinate role in the properties of the resultant nanofiller/polymer nanocomposite. The amount of nanofiller is expressed in terms of weight percentages (wt%). The different nanofiller materials used for the fabrication of polymer nanocomposites include clays, graphene, nanocellulose, carbon nanotubes (CNTs), etc.

4.2.1 NANOCLAY

Nanoclays are layered silicates, usually with a phyllosilicate or sheet structure, that contain metal oxides of alkali earth metals, alkali metals, calcium, and other metal oxides. Examples of clay-based nanofillers are montmorillonite (MMT), halloysite (Hal), bentonite, kaolinite, etc. Based on the arrangements of alumina and silica sheets, different types of clays are available. Among the several nanoclays, the most often used is MMT, which consists of two tetrahedral silica sheets fused to an octahedral sheet of alumina [8]. Based on their mineralogical composition, there are approximately 30 different types of nanoclays. Another type of nanoclay is Hal, which is a natural aluminosilicate clay with stoichiometry $Al_2Si_2O_5(OH)_4 \cdot nH_2O$. Nanoclay-based polymer nanocomposites have gained attention because the presence of nanoclay improves the mechanical and thermal properties of polymers [9] Also, it reinforces the structure of the polymer [10]. However, the homogeneous dispersion of the nanofiller and its chemical compatibility with the polymer is to be attained. This can be achieved through surface functionalization. Different methods available for the synthesis of polymer/nanoclay composites include the melt-blending method, solution-blending method, and *in situ* polymerization method [11]. The structure of polymer nanocomposites formed using layered nanoclays may be microcomposite, intercalated nanocomposite, or exfoliated nanocomposite [12].

4.2.2 CARBON NANOTUBES

Since its discovery by Lijima [13] in 1991, it has had profound applications in various fields of science and technology. Based on the method adopted for the preparation of CNTs, they are classified as multiwalled carbon nanotube (MWCNT) and single-walled carbon nanotube. Carbon in CNTs is sp^2 hybridized and it possesses high mechanical strength with very high Young's modulus and tensile strength. Pure CNTs are nonmagnetic whereas synthesized/functionalized CNTs are magnetic. The catalyst entangled in CNTs during its synthesis is responsible for its magnetic property.

CNTs have a promising role in the polymer matrix due to their unique mechanical, electrical, and thermal properties. The incorporation of CNTs results in the reinforcement of polymer and also makes it thermally conductive. However, the complexities involved in the fabrication of CNT/polymer nanocomposite include the difficulty in the dispersion of CNTs and poor interfacial interaction between CNTs and polymer matrix. The effective formation of CNT/polymer nanocomposites can be accomplished by surface functionalization of CNTs. Dispersion of CNTs into a polymer matrix can be achieved through mechanical processes such as ultrasonication, calendaring [14, 15], ball milling [16], and stir and extrusion method [17]. Various methods are available for the synthesis of CNTs.

4.2.3 GRAPHENE OXIDE

Graphene was discovered in 2004 by Geim and Novoselov [18]. Graphene has a high Young's modulus, tensile strength, thermal conductivity, electrical conductivity, and large surface area. These properties of graphene enable them to enhance the mechanical, electrical, and thermal properties of polymers [19, 20]. It was demonstrated that surface functionalization of graphene sheets (FGS) enhanced dispersion of polymers and also prevented the agglomeration of nanoparticles [20–22].

4.2.4 METAL NANOPARTICLES

Metal oxide nanoparticles have a large surface-to-volume ratio, optical properties, surface plasmon resonance, and quantum size effects. The methods available for the fabrication of polymer/metal oxide nanocomposites are: (1) *in situ* polymerization, (2) direct mixing/blending of metal oxide nanoparticles and polymer, and (3) sol–gel process. Polymers incorporated with metal oxide nanoparticles have applications in sensors, piezoelectric devices, fuel cells, and catalysts, etc. [21–23].

4.3 SURFACE FUNCTIONALIZATION/MODIFICATION METHODS OF NANOFILLERS

The surface functionalization of CNTs decreases its agglomeration and also increases the interfacial interactions between the CNT and the polymer. Figure 4.1 is a schematic representation of polymer nanocomposites without and with surface-treated nanoparticles [24].

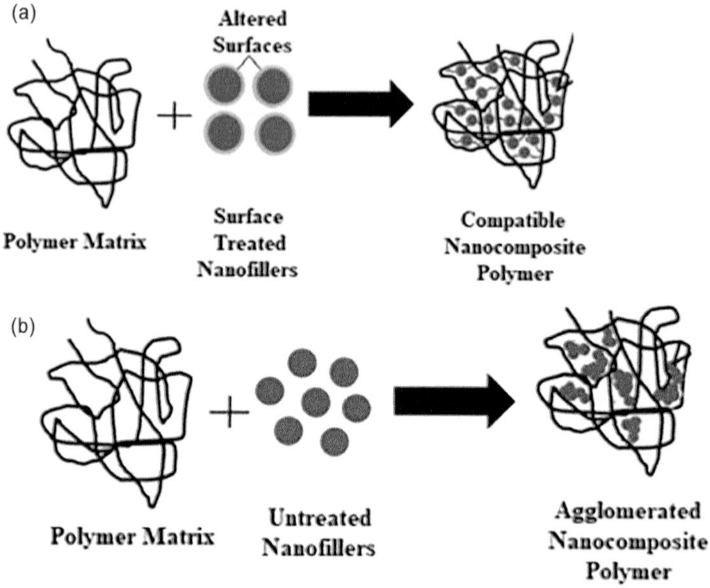

FIGURE 4.1 Schematic representation of polymer nanocomposites filled with (a) untreated nanomaterials and (b) surface-treated nanomaterials. (*Copyright permission*).
Source: Reprinted from Ref. [24]. https://creativecommons.org/licenses/by/4.0/

The methods adopted for surface modification can be broadly classified into covalent and noncovalent functionalization methods. Covalent functionalization involves the covalent linkage between functional groups on the nanotubes. Noncovalent functionalization is achieved through van der Waals' and π-stacking interactions, which are responsible for the noncovalent interaction between surface-active reagents and nanotubes. The mechanical and electrical conductivity of CNTs-based polymer nanocomposites can be improved by modifying the surface of CNTs with ionic liquids (ILs) [26–28]. The enhanced property is due to the cation–π, π–π, or van der Waals interaction between the imidazolium ring of ILs and CNTs. IL-modified CNTs have better dispersion than unmodified CNTs [29]. Functionalization of CNTs with various metal nanoparticles can be achieved [30–32]. These materials increased the nucleation and thermal properties, and also metal-decorated CNTs incorporated have specific applications in sensors, catalysis, etc. Hosseini et al. [33] reported the synthesis of TiO_2–Fe_3O_4–MWCNT hybrid nanotubes and it was incorporated into the PVDF polymer matrix.

The surface functionalization of graphene nanoparticles is also carried out. Zhao et al. [34] reported the FGS with poly(1-vinylimidazole). Also, the surface of graphene can be modified by the reduction of graphene-to-graphene oxide (rGO) [35], or by attaching hydroxy, carboxy, or epoxy groups to the graphene surface [36, 37].

4.4 SYNTHESIS OF SURFACE-FUNCTIONALIZED NANOFILLERS BASED FLUOROPOLYMER NANOCOMPOSITES

The synthesis of polymer nanocomposites can be carried out through (1) chemical or (2) mechanical processes. The different methods for the synthesis of polymer nanocomposite can be mainly classified into three: solution casting, melt blending, and *in situ* polymerization. Challenges involved in the synthesis of polymer nanocomposites are (1) compatibility between the fluoropolymer and nanoparticles, (2) homogeneous dispersion of nanofillers into the polymer matrix, and (3) agglomeration of nanoparticles. Surface modification of nanoparticles enhances their interaction with polymer matrices and also reduces the agglomeration of nanoparticles.

4.4.1 MELT INTERCALATION METHOD

This is a top-down approach and involves annealing a mixture of polymer and nanoparticles at molten temperature. This method enables the fabrication of polymer nanocomposites using polymers that are not suitable for *in situ* polymerization. The degradation of surface modification of nanofillers under high-temperature conditions is a major drawback of this method. The extent of degradation can be reduced either by optimizing the reaction temperature or by modifying the surface of the nanofiller with thermally stable materials.

Copper nanoparticles in a fluoropolymer matrix were synthesized by the melt intercalation method. The copper metal coated with poly(hexafluoropropylene [HFP] oxide, C_3F_6O) was annealed for the fabrication of copper nanoparticle-based fluoropolymer [38].

Li et al. [39] reported the synthesis of nanoclay (1.34 TCN, surface-modified montmorillonite) based poly(vinylidene fluoride) (PVDF) nanocomposites by the melt-intercalation method. PVDF nanocomposites were prepared via the melt-compounding method using a twin-screw extruder at 190°C with a screw speed of 180 rpm. Transmission microscopy images revealed that nanoclay particles are dispersed in the polymer matrix. As the concentration of nanoclay was increased, agglomeration of nanoclay took place. Also, nanoclay-modified PVDF has more β crystalline form and as the concentration of nanoclay is increased peak due to β-form in XRD increases.

Barium titanate ($BaTiO_3$) is a ceramic ferroelectric material with high relative permittivity. $BaTiO_3$ nanoparticles reinforced polymer matrices have high dielectric, thermal, and mechanical properties. These properties can also be enhanced by the surface modification of $BaTiO_3$ nanoparticles [40, 41]. Due to its high relative permittivity and low dielectric loss, $BaTiO_3$ nanowires (NWs) are considered to be a better nanofiller in polymer nanocomposites. Moreover, $BaTiO_3$-incorporated polymer nanocomposites have been reported to have widespread application in electrostatic capacitors, and energy storage devices, etc. [42].

Konno et al. reported the fabrication of fluoropolymers reinforced with $BaTiO_3$ nanoparticles [43]. $BaTiO_3$ polymer nanocomposite was synthesized by the sol–gel method. Uniform dispersion of $BaTiO_3$ NPs in *N*-methyl-2-pyrrolidinone (NMP) and PVDF in NMP was mixed together by stirring at 70°C for 12 h. Thin polymer nanocomposite films are obtained by spin coating the $BaTiO_3$ nanoparticles incorporated PVDF polymer on ITO substrates. Further synthesis of PVDF with the incorporation of surface-modified $BaTiO_3$ NPs was reported. Synthesized $BaTiO_3$ NPs were modified with fluoro silanes. The surface modification of $BaTiO_3$ enables the homogenous dispersion of NPs into polymer matrices.

Jiang et al. reported the synthesis of fluoro-polydopamine modified $BaTiO_3$ NWs (f-DOPA@$BaTiO_3$) incorporated PVDF-*co*-HFP p(VDF-HFP) fluoropolymer by solution blending method [44]. In this study, homogeneous dispersion of polymer-modified $BaTiO_3$ nanoparticles in DMF was obtained by ultrasonication, and it was then mixed with P(VDF-HFP) by stirring for 24 h. The resultant solution was cast on a glass plate. The polymer nanocomposite films were peeled off from the glass plate. Finally, it was dried for 12 h at 40°C.

Synthesis of a multifunctional nanohybrid PVDF polymer nanocomposite was synthesized by solution mixing and electrospinning method [45]. Magnetic Fe_3O_4 nanoparticles and conductive graphene oxide (GO) are incorporated into the PVDF matrix. Electrospinning was carried out at a voltage of 12 kV between the spinneret and collector with a working distance of 15 cm.

Céspedes et al. employed plasma polymerization for modifying the surface of sodium montmorillonite nano clay (Na^+-MMT) with methyl methacrylate (MMA) and styrene (St) as monomers [46]. The greater interfacial interaction and dispersion of the Na^+-MMT nanoclay modified by styrene plasma in the polystyrene (PS) matrix gave way to an improvement in thermal stability as well as an increase in 34% of Young's modulus for the PS/Na^+-MMT/St. The temperature of degradation for the PS/Na^+–MMT/St nanocomposite was 402.9°C, for PS/Na^+–MMT/MMA nanocomposite was 387.5°C, and for the PS it was 377.5°C. The result indicates that the incorporation of the nanoclay and its dispersion by mixing gives thermal stability to the PS, but a surface modification of Na^+-MMT with styrene plasma presents a better interaction with the PS matrix, since both present the same functional group (St) that interact within themselves, provoking better compatibility between both phases, promoting an improvement in the dispersion of the analyzed sample.

Poly(tetrafluoroethylene)/graphene nanocomposites (PTFE/GNs) was prepared by electrostatic self-assembly method. Polyethyleneimine (PEI)-modified PTFE polymers are positively charged, whereas the GO is negatively charged. When the solution of GO and PEI-coated PTFE are mixed together electrostatic attraction occurs resulting in the formation of a PTFE–GO hybrid. The fabrication of PTFE-GN was done by cold briquetting and hot sintering of PTFE–GN hybrids [47].

Two polymer/GN were mainly prepared via solution mixing and melt mixing methods [48]. Poly (MMA) functionalized graphene/poly(vinylidene fluoride) nanocomposites were synthesized by atom transfer radical polymerization. An isolated single layer of reduced GO (RGO) enhances the thermal and electrical properties at very low percolation values. Powdered GO in DMF is dispersed to the PVDF/DMF solution through ultrasonication. GO/PVDF polymer nanocomposite film is obtained on a glass substrate and the resultant PVDF/GO film is subjected to a hot press at 150°C for 2 h to reduce the GO to RGO. This method fabricated a PVDF polymer matrix with functionalized GO. The first step is the surface modification of GO with polypyrrole (PPy). Finally, PVDF-based composites were prepared by ultrasonication.

4.5 PROPERTIES OF SURFACE FUNCTIONALIZED NANOFILLERS-BASED FLUOROPOLYMER NANOCOMPOSITES

4.5.1 PIEZOELECTRIC PROPERTY

The properties of surface-functionalized nanofiller-incorporated fluoropolymers depend on the size, specific surface area, molecular structure, nanofiller employed, and type of surface functionalization. The properties of fluoropolymers can be tailored by the functionalization of the surface of nanofillers. The desired properties of electronic devices are lightweight, high conductivity, high dielectric permittivity, low dielectric loss, and high breakdown strength. The discovery of piezoelectricity in semicrystalline PVDF by Kawai [49] was a major breakthrough in the development of energy devices. Even though fluoropolymers show the low dielectric property, enhanced dielectric permittivity, and minimum dielectric loss can be achieved by incorporating nanoparticles in the polymer matrix.

Chung et al. [48] have studied the effect of adding reduced GO (RGO) to PVDF. The study revealed that there is a remarkable change in piezoelectric and ferroelectric properties of pure PVDF and PVDF/RGO polymer nanocomposite. The optimum concentration of RGO was at 0.3 wt%. At this concentration β phase (100%) and high normalized strain, Smax/Emax of 16.66 pm/V was achieved. Also, PVDF/RGO polymer nanocomposite exhibited an increase in dielectric constant (320%) and remnant polarization (200%). The improved performance of PVDF/RGO polymer nanocomposite helped to be a better energy storage device. Zuo–Wan et al. reported that the dielectric properties of PVDF were enhanced when GO was functionalized with PPy [50].

Dielectric constant of a PVDF matrix with the inclusion of ($CaCu_3Ti_4O_{12}$ [CCTO]) and CNTS was studied [51]. It was found that the dielectric constant of CCTO/PVDF is 6.5 with a dielectric loss of 0.075, whereas when CNTS were added to the CCTO/PVDF matrix the dielectric constant increased to 55.733 with a dielectric loss of 0.251. Also, the energy storage density of CCTO/PVDF polymer matrix was 0.617 J cm^{-3} and that of MWCNT/CCTO/PVDF was 1.395 J cm^{-3}. When CCTO/PVDF was modified with Ag nanoparticles, the dielectric constant increased to 110 [52]. The enhancement of dielectric properties with the incorporation of conducting nanofibers can be attributed to the Maxwell-Wagner-Sillars effect [53].

The incorporation of polymer-modified $BaTiO_3$ nanoparticles into fluoropolymer increased the dielectric constant (*k*) and reduced breakdown strength. The properties of fluoropolymer vary with the concentration of $BaTiO_3$ nanoparticles. Compared to the pure polymer (13.75 J cm^{-3}), the energy storage densities of the nanocomposite films with 2.5 vol% and 5 vol% f-DOPA@$BaTiO_3$ NWs were reported to be 21.35 J cm^{-3} and 22.58 J cm^{-3}. It was observed that the incorporation of 30 vol% $BaTiO_3$ into the PVDF matrix improved its dielectric constant to 32 and reduced dielectric loss to 0.05 [42]. Whereas when $BaTiO_3$ was modified with 3,3,3-trifluoropropyl methyl silane and resultant NPs were incorporated into PVDF matrix the dielectric constant increased to 44 [54]. Optical microscope images revealed the presence of micron-sized agglomerates in a polymer matrix with unmodified $BaTiO_3$ nanoparticles, and little agglomerates were observed when the surface of $BaTiO_3$ nanoparticles was modified with alkyl silane.

PVDF was the first fluoropolymer to exhibit piezoelectric properties [55]. However, the piezoelectric coefficients of polymers are less. This can be enhanced by the addition of nanofillers. In the case of PVDF, the β phase is polar and is responsible for the piezoelectric, pyroelectric, and ferroelectric properties of PVDF. The enhancement of the β phase can be achieved through the incorporation of nanofillers into PVDF.

Wu et al. reported that the addition of CNT to PVDF increased the β phase concentration to 88% [56]. The piezoelectric coefficient of unmodified PVDF was 16.8, and that of CNT/PVDF was 31.3. The piezoelectric property of fluoropolymer nanocomposites makes it a good candidate for biological applications such as gene delivery, drug delivery, etc. Wang et al. [57] proved that PVDF/nanoclay nanofibers fabricated through the electrospinning method act as a piezoelectric sensor for monitoring pulse and respiratory rate. PVDF/nanoclay fiber sensors record the dynamic force performed by a finger, and responses were measured by an oscilloscope.

The dispersion of nanofillers into fluoropolymer matrices significantly enhances its mechanical strength. The mechanical property depends upon the nature and amount of the nanofiller. Yan et al. [58] studied the mechanical strength of PTFE/nano-EG composites reinforced with different nanoparticles. It was observed that pure PTFE and PTFE with SiO_2 and TiO_2 have no significant yield strength. Compared to pure PTFE, nanoparticle-incorporated PTFEs have enhanced mechanical properties. Out of different nanocomposites, nano-Al_2O_3 incorporated poly nanocomposite has maximum tensile strength, and nano-SiO_2/PTFE composite has maximum elastic modulus.

4.6 APPLICATION OF NANOFILLER-MODIFIED FLUOROPOLYMER IN ENERGY DEVICES

The special chemical and physical properties of fluoropolymers, such as chemical resistance, thermal stability, cryogenic properties, low coefficient of friction, low surface energy, low dielectric constant, high volume, surface resistivity, and flame resistance, make them applicable in various fields. They have found profound applications in chemical, electronic, and electrical, space, construction, architectural, food processing, medical, and automotive industries. Pure fluoropolymers suffer from low dielectric constant. Different strategies were developed to improve the efficiency of fluoropolymers as energy devices. Among all the methods, the incorporation of nanoparticles into fluoropolymers is outstanding. Table 4.1 includes the list of nanoparticles incorporated in the PVDF matrix as piezoelectric devices.

TABLE 4.1 List of Nanoparticles Incorporated PVDF Matrix as Piezoelectric Devices

Sl. No.	Nanocomposite Modified Fluoropolymer	Significant Improvement in Properties	References
1	ZnO incorporated PVDF ZnO/PVDF composite	Enhance dielectric constant and show a lower percolation threshold	[59]
2	Multiwalled nanotube incorporated (PVDF/MWCNT) nanocomposite films	The dielectric permittivity (550) of nanocomposites increases with increasing loading of MWCNT (5.99 % MWCNT) at relatively low dielectric losses (0.4) and conductivity levels	[60]
3	(i) Reduced graphene oxide (rGO) incorporated PVDF nanocomposite and (ii) poly (vinyl alcohol)-modified rGO incorporated PVDF nanocomposite	Compared to rGO/PVDF composites, higher dielectric constant and lower loss factor were simultaneously achieved in rGO-PVA/PVDF nanocomposites	[61]
4	Strontium titanate ($SrTiO_3$) nanowires incorporated PVDF composites	Enhanced dielectric constant and reduced dielectric loss at a low filler content	[62]
5	Polyvinylidene fluoride (PVDF)-based nanocomposites with graphene and ionic liquid-modified graphene as fillers	Higher dielectric permittivity together with a lower loss factor is achieved	[63]
6	Polydopamine (PDA) coated carbon nanotube (CNT)/poly(vinylidene fluoride) (PVDF)	Enhanced dielectric permittivity and reduced dielectric loss	[64]

TABLE 4.1 *(Continued)*

Sl. No.	Nanocomposite Modified Fluoropolymer	Significant Improvement in Properties	References
7	Silver nanoparticles incorporated poly(vinylidene fluoride) (Ag/PVDF)	A greatly enhanced dielectric permittivity (257.2) and low loss tangent (0.26)	[65]
8	γ-Oxopyrenebutyric acid functionalized graphene and $BaTiO_3$ nanofillers incorporated PVDF composite	Enhanced dielectric constant and low dielectric loss	[66]
9	Graphite nanosheets/PVDF composite	Enhanced dielectric constant, increased breakdown field strength, higher energy storage density, and lower dielectric loss	[67]
10	Ag nanomaterials deposited reduced graphene oxide (rGO) hybrid particles (Ag-rGO HPs) incorporated PVDF nanocomposites (Ag-rGO HPs/PVDF composites)	Enhanced dielectric constant	[68]

Nanofiller-incorporated fluoropolymers are employed in electronics due to their unique electrical conductivity, charge storage, and electromagnetic interference shielding. The actuation properties of PVDF, CNT/PVF, and carbonic acid-functionalized CNT/PVDF were compared. It was observed that carbonic acid-functionalized CNT/PVDF has a higher saturation current and minimum polarization resistance, which makes it potentially applicable in sensing. Poly(dopamine) modified $BaTiO_3$ NWs incorporated PVDF polymers have been used in energy storage devices. The reported fluoropolymer has an excellent dielectric constant and could maintain the reduction of breakdown strength. The fluoropolymer with 5 vol% f-DOPA@$BaTiO_3$ NWs possesses 12.87 J cm^{-3} of energy density proving its utility as a promising energy storage device. TiO_2-incorporated PVDF nanocomposite acts as water treatment [69–71]. Even though the physical, chemical, and mechanical properties of polymeric PVDF favor its use in ultrafiltration, proteins, oils, and natural organic matter present in water cause fouling of PVDF membranes. The fouling effect can be reduced by surface modification of PVDF using nanomaterials. Good stability, cheapness, hydrophilicity, and nontoxicity make TiO_2 a good candidate for PVDF membrane modification. Meldrum's acid-modified CNTs/PVDF membrane has proven to be an excellent eco-friendly adsorbent material for toxic pollutants. The membrane also eliminates Fe_2O_3 nanoparticles and crystal violet from water.

The piezoelectric property of nanofiller-incorporated fluoropolymers enabled them to be used in biomedical monitoring applications. Simchi et al. fabricated microporous zirconium-based metal-organic frameworks that incorporated PVDF nanocomposite which is used as a wearable piezoelectric device for monitoring pulse [72]. A wearable Bi_2S_3/PVDF/PPy polymer nanocomposite pressure sensor was successfully applied for monitoring the pulse rate of human radial arteries [73].

Silver nanoparticles stabilized with aloe vera extract (Ag@Ale–Nps) were used as nanofillers to enhance the dielectric constant of PVDF. The increased dielectric constant of nanoparticle-incorporated PVDF can be attributed to the hydroxyl and carboxyl groups on the surface of AgNPs [74]. Zhang et al. [25] reported the preparation of 1D nano-conductive filler/PVDF dielectric nanocomposite using natural sepiolite as the core of nanofiber and conductive polymer PPy as the shell of nanofiber. The effect of the modified layer on the dielectric constant of fluoropolymer and the possible polarization mechanism is explained in Figure 4.2.

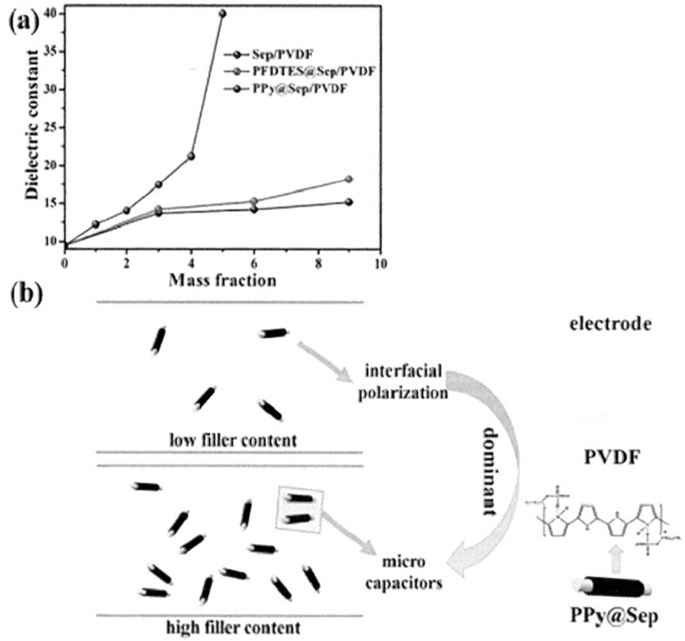

FIGURE 4.2 (a) Dielectric constants of Sep/PVDF, PFDTES@Sep/PVDF and PPy@Sep/PVDF composites at 100 Hz. (b) Schematic diagram of the polarization mechanism of PPy@Sep/PVDF nanocomposites.

Source: Reprinted with permission from Zhang et al. (2021). Copyright 2021. Elsevier.

4.7 CONCLUSIONS

Science and technology work together with nanotechnology to advance society. The application of nanotechnology to various scientific disciplines has increased the effectiveness of current systems. The chemical and physical properties of a system are improved as a result of the integration of nanoparticles. This then makes it more useful in numerous fields. Fluoropolymers' use of nanoparticles contributed to the creation of cutting-edge materials with academic and commercial significance. By altering the nanofillers and functionalizing the nanofiller surfaces, fluoropolymers' characteristics can be customized. This chapter includes a quick overview of several nanofillers and their modifications. Additionally, the creation of fluoropolymers with nanofiller inclusions and their uses are covered.

KEYWORDS

- **fluoropolymers**
- **functionalized nanofillers**
- **functionalized fluoropolymer nanocomposites**
- **synthesis**
- **properties**
- **applications**

REFERENCES

1. Ebnesajjad, S. Fluoroplastics: Non-melt processible fluoroplastics. *Plastics Design Library* 2000.
2. Plunkett, R.J. Tetrafluoroethylene polymers. U.S. Patent 2, 230, 654, February 4, 1941.
3. Ebnesajjad, S. *Fluoroplastics, Melt Processible Fluoropolymers*. Volume 2: Plastics Design Library, 2003.
4. McKeen, L.W. *Fatigue and Tribological Properties of Plastics and Elastomers*. 2nd edn., William Andrew Publishing, 2010.
5. Hamdy, A.S. Fluoropolymer coatings for corrosion protection in highly aggressive environments, in: *High-Performance Organic Coatings*, Woodhead Publishing, 2008.

6. Al-Haj Ali, M., Elleithy, R.H. Viscoelastic properties of polypropylene/organo-clay nano-composites prepared using miniature lab mixing extruder from masterbatch. *Journal of Applied Polymer Science* 2011, 121, 27–36.
7. Hussain, F., Hojjati, M., Okamoto, M., Gorga, R.E. Polymer-matrix nanocomposites, processing, manufacturing, and application: An overview. *Journal of Composite Materials* 2006, 40, 1511–1575.
8. Chen, X., Gonsalves, K.E. Synthesis and properties of an aluminum nitride/polyimide nanocomposite prepared by a nonaqueous suspension process. *Journal of Materials Research* 1997, 12, 1274–1286.
9. Majeed, K., Jawaid, M., Hassan, A., Abu Bakar, A., Abdul Khalil, H.P.S., Salema, A.A., Inuwa, I. Potential materials for food packaging from nanoclay/natural fibres filled hybrid composites. *Master of Design* 2013, 46, 391–410.
10. Osman, M.A., Atallah, A. High-density polyethylene micro and nanocomposites: Effect of particle shape, size and surface treatment on polymer crystallinity and gas permeability. *Macromolecular Rapid Communications* 2004, 25(17), 1540–1544.
11. Liu, P. Polymer modified clay minerals: A review. *Applied Clay Science* 2007, 38, 64–76.
12. Jawaid, M., Qaiss, A.K., Bouhfid, R. *Nanoclay Reinforced Polymer Composites: Nanocomposites and Bionanocomposites*. Springer: Singapore, 2016.
13. Lijima, S. Synthesis of carbon nanotubes. *Nature* 1991, 354, 56
14. Alexandre, M., Dubois, P. Polymer-layered silicate nanocomposites: Preparation, properties and uses of a new class of materials. *Materials Science and Engineering R: Reports* 2000, 28, 1–63.
15. Gojny, F.H., Wichmann, M.H.G., Köpke, U., Fiedler, B., Schulte, K. Carbon nanotube reinforced epoxy-composites: Enhanced stiffness and fracture toughness at low nanotube content. *Composites Science and Technology* 2004, 64, 2363–2371.
16. Thostenson, E.T., Chou, T.W. Processing-structure-multi-functional property relationship in carbon nanotube/epoxy composites. *Carbon* 2006, 44, 3022–3029.
17. Sandler, J.K.W., Shaffer, M.S.P., Prasse, T., Bauhofer, W., Schulte, K., Windle, A.H. Development of a dispersion process for carbon nanotubes in an epoxy matrix and the resulting electrical properties. *Polymer* 1999, 40, 5967–5971.
18. Geim, A.K., Novoselov, K.S. The rise of graphene. *Nature Materials* 2007, 6, 183.
19. Ma, P.C., Tang, B.Z., Kim, J.K. Converting semiconducting behavior of carbon nanotubes using ball milling. *Chemical Physics Letters* 2008, 458, 166–169.
20. Wan, C., Chen, B. Reinforcement and interphase of polymer/graphene oxide nanocomposites. *Journal of Materials Chemistry* 2012, 22, 3637–3646.
21. Tan, B., Thomas, N.L. A review of the water barrier properties of polymer/clay and polymer/grapheme nanocomposites 2016, 514, 595–612.
22. Mukhopadhyay, P., Gupta, R.K. Graphite, graphene and their polymer nanocomposites, 1st edn., CRC Press, 2013, p. 589.
23. Pradeep, T., Anshup. Noble metal nanoparticles for water purification: A critical review. *Thin Solid Films* 2009, 517, 6441–6778.
24. Saman, N.M., Ahmad, M.H., Buntat, Z. Application of cold plasma in nanofillers surface modification for enhancement of insulation characteristics of polymer nanocomposites: A review. *JEE Access* 2021, 9, 80906–80930.

25. Chen, S., Qiao, R., Liu, Z., Zhang, D. Enhanced dielectric constant of PVDF-based nanocomposites with one-dimensional core-shell polypyrrole/sepiolite nanofibers. *Composites Part A: Applied Science* 2021, 145, 106384–106389.
26. Miehr, R., Tratnyek, P.G., Bandstra, J.Z., Scherer, M.M., Alowitz, M., Bylaska, E.J. The diversity of contaminant reduction reactions by zero-valent iron: Role of the reductate. *Environmental Science & Technology* 2004, 38, 139–147.
27. Wang, Y., Tang, W., Yang, Y., Xie, X. Progress in imidazolium ionic liquids assisted fabrication of carbon nanotube and graphene polymer composites. *Polymer* 2013, 5, 847–872.
28. Feng, J., Sun, M., Li, L., Wang, X., Duan, H., Luo, C. Multiwalled carbon nanotubes-doped polymeric ionic liquids coating for multiple headspace solid-phase microextraction. *Talanta* 2014, 123, 18–24.
29. Subramaniam, K., Das, A., Heinrich, G. Development of conducting polychloroprene rubber using imidazolium based ionic liquid modified multi-walled carbon nanotubes. *Composites Science and Technology* 2011, 71, 1441–1449.
30. Tunckol, M., Durand, J., Serp, P. Carbon nanomaterial–ionic liquid hybrids. *Carbon* 2012, 50, 4303–4334.
31. Dobrzański, L., Pawlyta, M., Krztoń, A., Liszka, B., Labisz, K. Synthesis and characterization of carbon nanotubes decorated with platinum nanoparticles. *Journal of Achievements in Materials and Manufacturing Engineering* 2010, 39, 184–189.
32. Sharma, P., Sharma, R.K. Platinum functionalized multiwall carbon nanotube composites as recyclable catalyst for highly efficient asymmetric hydrogenation of methyl pyruvate. *RSC Advances* 2015, 5, 102481–102487.
33. Samadi, A. Ahmadi, R., Hosseini, S.M. Influence of TiO2-Fe3O4-MWCNT hybrid nanotubes on piezoelectric and electromagnetic wave absorption properties of electrospun PVDF nanocomposites. *Organic Electronics* 2019, 75, 105405–105412.
34. Zhao, W., Tang, Y., Xi, J., Kong, J. Functionalized graphene sheets with poly(ionic liquid)s and high adsorption capacity of anionic dyes. *Applications of Surface Science* 2015, 326, 276–284.
35. Shin, D.G., Yeo, H., Ku, B., Goh, M., You, N. A facile synthesis method for highly water-dispersible reduced graphene oxide based on covalently linked pyridinium salt. *Carbon* 2017, 121, 17–24.
36. Cai, L., Lin, Z., Wang, M., Pan, F., Chen, J., Wang, Y., Shen, X., Chai, Y. Improved interfacial H_2O supply by surface hydroxyl groups for enhanced alkaline hydrogen evolution. *Journal of Materials Chemistry A* 2017, 5, 24091–24097.
37. Vinod, S., Tiwary, C.S., Samanta, A., Ozden, S., Narayanan, T.N., Vajtai, R., Agarwal, V., Singh, A.K., John, G., Ajayan, P.M. Graphene oxide epoxy (GO-xy): GO as epoxy adhesive by interfacial reaction of functionalities. *Advanced Materials Interfaces* 2018, 5, 1700657.
38. Safonov, A., Sulyaeva, V., Timoshenko, N., Starinskiy, S. Synthesis of copper nanoparticles in a fluoropolymer matrix by annealing in vacuum. *Physics Letters A* 2017, 381, 2013–2016.
39. Peng, Q., Peihong, C., Xujun, L., Tianxi, L., Shu-Chen, H., Li, T. The preparation of PVDF/clay nanocomposites and the investigation of their tribological properties. *Wear* 2017, 266, 713–720.

40. Kim, H., Fernado, T., Li, M., Lin, Y., Tseng, T.L. Fabrication and characterization of 3D printed $BaTiO_3$/PVDF nanocomposites. *Journal of Materials* 2017, 52, 197–206.
41. Tavangar, M., Heidari, F., Hayati, R., Tabatabaei, F., Vashaee, D., Tayebi, L. Manufacturing and characterization of mechanical, biological and dielectric properties of hydroxyapatite-barium titanate nanocomposite scaffolds. *Ceramics International* 2020, 46, 9086–9095.
42. Yu, K., Wang, H., Zhou, Y., Bai, Y., Niu, Y. Enhanced dielectric properties of $BaTiO_3$/poly(vinylidene fluoride) nanocomposites for energy storage applications. *Journal of Applied Physics* 2013, 113, 034105.
43. Kobayashi, Y., Tanase, T., Tabata, T., Miwa, T., Konno, M. Fabrication and dielectric properties of the $BaTiO_3$–polymer nano-composite thin films. *Journal of the European Ceramic Society* 2008, 28(1), 117–122.
44. Wang, G., Huang, X., Jiang, P. Bio-inspired fluoro-polydo pamine meets barium titanate nanowires: A perfect combination to enhance energy storage capability of polymer nanocomposites. *ACS Applied Materials & Interfaces* 2017, 9, 7547–7555.
45. Samadi, A., Hosseini, S.M., Mohseni, M. Investigation of the electromagnetic microwaves absorption and piezoelectric properties of electrospun Fe_3O_4-GO/PVDF hybrid nanocomposites. *Organic Electronics* 2018, 59, 149–155.
46. Zare, Y., Garmabi, H. Thickness, modulus and strength of interphase in clay/polymer nanocomposites. *Applied Clay Science* 2015, 105–106, 66–70.
47. Jiang, H., Chen, L., Chai, S., Yao, X., Chen, F., Fu, Q. Facile fabrication of poly (tetrafluoroethylene)/graphene nanocomposite via electrostatic self-assembly approach. *Composites Science and Technology* 2014, 103, 28–35.
48. Ataur, R., Chung, M., Gwiy, S. Synthesis of PVDF-graphene nanocomposites and their properties. *Journal of Alloys and Compounds* 2013, 581, 724–730.
49. Kawai, H. *Japanese Journal of Applied Physics* 1969, 8, 975.
50. Zhen-zhen, H., Xu, Y., Jing-hui, Y., Nan, Z., Ting, H., Yong, W., Zuo-wan, Z. Largely enhanced dielectric properties of poly(vinylidene fluoride) composites achieved by adding polypyrrole-decorated graphene oxide. *Composites Part A: Applied Science* 2018, 104, 89–100.
51. Lu, J., Wong, C.P. Recent advances in high-*k* nanocomposite materials for embedded capacitor applications. *IEEE Transactions on Dielectrics and Electrical Insulation* 2008, 15, 1322–1328.
52. Dang, Z.M., Yuan, J., Zha, J.W., Zhou, T., Li, S.T., Hu, G.H. Fundamentals, processes and applications of high-permittivity polymer–matrix composites. *Progress in Materials Science* 2012, 57, 660–723.
53. Nan, C.W., Shen, Y., Ma, J. Physical properties of composites near percolation. *Annual Review of Materials Research* 2010, 40, 131–151.
54. Kawai, H. The piezoelectricity of poly (vinylidene fluoride). *Japanese Journal of Applied Physics* 1969, 8, 975–976.
55. Wu, C.M., Chou, M.H. Polymorphism, piezoelectricity and sound absorption of electrospun PVDF membranes with and without carbon nanotubes. *Composites Science and Technology* 2016, 127, 127–133.
56. Xin, Y., Qi, X., Tian, H., Guo, C., Li, X., Lin, J., Wang, C. Full-fiber piezoelectric sensor by straight PVDF/nanoclay nanofibers. *Materials Letters* 2015, 164, 136–139.

57. Soumyaranjan, M., Kumaran, K.T., Sivakumaran, R., Paul Pandian, S., Subrata, K. Synthesis of PVDF/CNT and their functionalized composites for studying their electrical properties to analyze their applicability in actuation & sensing. *Colloids and Surfaces A: Physicochemical and Engineering Aspects* 2016, 509, 684–696.
58. Yanhong, Y., Zhining, J., Yulin, Y. *Procedia Environmental Sciences* 2011, 10, 929–935.
59. Zhu, J., Ji, X., Yin, M., Guo, S., Shen, J. Poly (vinylidene fluoride) based percolative dielectrics with tunable coating of polydopamine on carbon nanotubes: Toward high permittivity and low dielectric loss. *Composites Science and Technology* 2017, 144, 79–88.
60. Shaonan, C., Sheng, C., Rui, Q., Haoran, X., Zijin, L., Hang, L., Dou, Z. Enhanced dielectric constant of PVDF-based nanocomposites with one-dimensional core-shell polypyrrole/sepiolite nanofibers. *Composites Part A: Applied Science and Manufacturing* 2021, 145, 106384.
61. Wang, D., Bao, Y., Zha, J.W., Zhao, J., Dang, Z.M., Hu, G.H. Improved dielectric properties of nanocomposites based on poly(vinylidene fluoride) and poly(vinyl alcohol)-functionalized graphene. *ACS Applied Materials & Interfaces* 2012, 4, 6273–6279.
62. Dan, Z., Jiang, J., Zhang, X., Qian, J., Shen, Z., Nan, C., Shen, Y. Interfacial effects of $BaTiO_3$@TiO_2 nanofibers on dielectric relaxation processes of P(VDF-TrFE-CFE) nanocomposites. *Ceramics International* 2020, 46, 1119–1123.
63. Tuichai, W., Karaphun, A., Ruttanapun, C. Improved dielectric properties of PVDF polymer composites filled with Ag nanomaterial deposited reduced graphene oxide (rGO) hybrid particles. *Materials Research Bulletin* 2022, 145, 111552.
64. Li, X., Wang, X., Weng, L., Yu, Y., Zhang, X., Liu, L., Wang, C. Dielectrical properties of graphite nanosheets/PVDF composites regulated by coupling agent. *Materials Today Communications* 2020, 21, 100705.
65. Xu, P., Cui, Z., Chen, H., Ding, Y. Synergistic enhanced dielectric properties of PVDF nanocomposites containing γ-oxo-pyrenebutyric acid functionalized graphene and $BaTiO_3$ nanofillers. *Composites Communications* 2019, 13, 63–69.
66. Liu, S., Zhai, J. Improving the dielectric constant and energy density of poly(vinylidene fluoride) composites induced by surface-modified $SrTiO_3$ nanofibers by polyvinylprrolidone. *Journal of Materials Chemistry A* 2014, 3, 1511–1517.
67. Wang, G., Yuan, D., Yan, X., Lin, G. Fabrication of radial ZnO nanowire clusters and radial ZnO/PVDF composites with enhanced dielectric properties. *Advanced Functional Materials* 2010, 18, 2584–2592.
68. Xu, P., Gui, H., Wang, X., Hu, Y., Ding, Y. Improved dielectric properties of nanocomposites based on polyvinylidene fluoride and ionic liquid-functionalized graphene. *Composites Science and Technology* 2015, 117, 282–288.
69. Rahimpour, A., Jahanshahi, M., Rajaeian, B., Rahimnejad, M. TiO_2 entrapped nano-composite PVDF/SPES membranes: Preparation, characterization, antifouling and antibacterial properties. *Desalination* 2011, 278, 343–353.
70. Boributh, S., Chanachai, A., Jiraratananon, R., Modification of PVDF membrane by chitosan solution for reducing protein fouling. *Journal of Membrane Science* 2009, 342, 97–104.

71. Liang, S., Xiao, K., Mo, Y., Huang, X. A novel ZnO nanoparticle blended polyvinylidene fluoride membrane for anti-irreversible fouling. *Journal of Membrane Science* 2012, 394, 184–192.
72. Bentolhoda, H.M., Mahdi, H., Abdolreza, S. Self-powered wearable piezoelectric sensors based on polymer nanofiber–metal–organic framework nanoparticle composites for arterial pulse monitoring. *ACS Applied Nano Materials* 2020, 3, 8742–8752.
73. Sushmitha, V., Sushmee, B. Bi_2S_3/PVDF/Ppy-Based freestanding, wearable, transient nanomembrane for ultrasensitive pressure, strain, and temperature sensing. *ACS Applied Bio Materials* 2021, 4, 14–23.
74. Phromviyo, N., Chanlek, N., Thongbai, P., Maensiri, S. Enhanced dielectric permittivity with retaining low loss in poly(vinylidene fluoride) by incorporating with Ag nanoparticles synthesized via hydrothermal method. *Applications of Surface Science* 2018, 446, 59–65.

CHAPTER 5

Application of High Energy Materials for Propellants, Explosives, and Pyrotechnics

TRINATH BISWAL

Department of Chemistry, Veer Surendra Sai University of Technology, Burla, Orissa, India

ABSTRACT

Materials required for defense sectors must possess some unique properties such as high strength, lightweight, durability, good impact strength, stability, and cost-effectiveness. Because of their superior performance, high-energy materials (HEMs) are commonly used for this purpose. Because of their specific multidisciplinary properties, HEMs are best suited for this sector. The use of HEMs in various fields includes rocket propellants, pyrotechnics for celebrations, blasting materials, military explosives, and even some HEMs used in construction. Propellants, explosives, and pyrotechnics cause extensive degradation of the environment; therefore, researchers are focusing on eco-friendly, sustainable materials that can replace traditional HEMs. For improving the performance of HEMs, it is highly essential to accelerate the oxygen balance to maintain stoichiometry. In this study, emphasis is given to the development of new, improved high-performance explosives, propellants, and pyrotechnics, their properties, and their applications.

Advances in Energy Materials: Materials Fabrication, Properties and Applications.
Lishin T, Maciej Jaroszewski, and Sabu Thomas (Eds.)

DOI: 10.1201/9781779643674-5

5.1 INTRODUCTION

High-energy materials (HEMs) possess the capability of generating an excessive quantity of energy via chemical reactions through a self-sustaining pathway, where the reaction rate is much quicker and, in some cases, proceeds in a swift manner (taking a fraction of a second). Such attributes of modern materials can store an excessive amount of chemical energy, which can be further used effectively in numerous important fields, especially in space defense, aerospace, and satellites as propellants, explosives, and pyrotechnics. HEMs contain both oxidizing groups (-NO_2) and reducing groups, or both. These materials are normally termed as oxidizer, which is the combination of fuel elements like gun powder (KNO_3 + C + S). These categories of valuable materials undergo stimulation through mechanical, thermal, and electrical devices [1, 2]. Such materials can cause rapid degradation with the production of excessive quantities of heat, noise, light, and other forms of energy, along with the liberation of a substantial volume of toxic gaseous materials. The generation of energy from HEMs normally depends upon the nature and kinds of materials, their density, chemical composition, structure, ease of decomposition, heat or enthalpy of formation and decomposition, etc. HEMs of various categories, such as explosives, propellants, and fireworks, whose properties depend upon their design and use [3]. The burning of wood, coal, and wax generates energy and some gaseous components, but it is a relatively slow process; however, the combustion of HEMs generates an excessive quantity of energy and the rate of combustion is extremely high. The quick release of a huge amount of energy at a very rapid rate is the reason for its increasing application in many important devices, mostly for the manufacturing of explosives and propellants. A new interesting composite propellant termed as hydoxy terminated polybutadiene (HTPB) was developed by using Al metal as fuel and ammonium perchlorate (oxidizer), which shows high performance in rocket engines, propellant missiles, and space vehicles. These propellants are much more effective (Isp 250–260 s) than any traditional propellants like nitrocellulose (NC), nitroglycerine (NG), etc. The composite propellant (HTPB) exhibits better strain ability because of a lower glass transition temperature (T_g) [4, 5]. The most important use of this kind of propellant is its effective application in subzero temperatures. Throughout the world, SNPE, Royal, and Thiokol Ordnance are the major manufacturers of these propellants. The novelty of this material is that it can be used both for civil and military operations. The properties of the materials are best suited for propellants, explosives, and fireworks. The propellants are rich in fuel and have recently been developed into new categories of improved propellants where their performance is much superior owing to the generation of huge amounts of energy. Scheme 5.1 represents the structure of NG and NC [6]. The importance of this chapter is that it provides knowledge regarding the design and development of materials suitable for military operations. Propellants are highly important in space applplication, because solid propelllants are used as rocket fuel. The other imporatnt use of propelants are airbrushing and pressure washing. Sometines propellant can be used as gas generator that can be used for gas supply toopearte turbopumps of liquid rockets. HEMs are used for syntehsis of solid rocket propellants, explosives, and powder gun propellants and is highly important because basically used in weaponry, and space rocket industry. It provides superior perfermance as compared with conventional NG and NC.

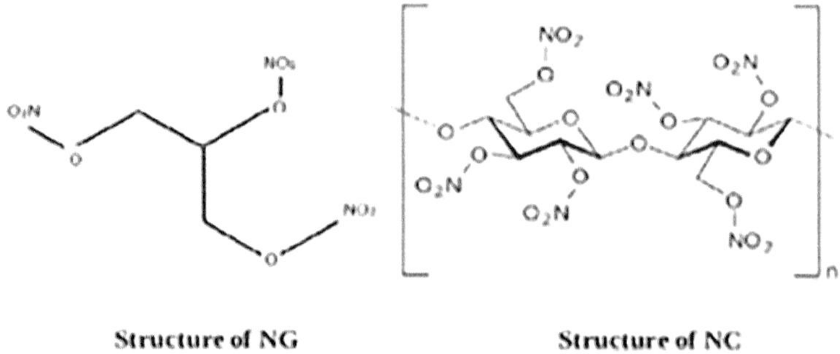

SCHEME 5.1 Structure of NG and NC.

Now a days, the fuel-rich propellants based on Mg are in craze, but the boron-based solid propellants are more preferred among all the metals listed as fuel-rich componets because of their extra energy potential. But the major demerit of these materials is the deficiency of effective pyrolyzability of boron and the need to increase the pyrolizability. Hence, some suitable energetic binders and additives are added to it to solve this problem. Gun propellants are usually a triple-base conventional propellant, and the three major components of them are NG, NC, and picrite (nitrguanidine [NQ]). These propellants now occupy a key position in field gun ammunition and tanks because of their low flame temperature, low barrel erosion, flash, less combustion, long shelf life, etc. For the environmental point of view, attention has always been paid to replacing the NQ with the addition of energetic plasticizers and oxidizers. In high explosive material sectors, the weapon designers and developers are facing many challenges, such as shaped charges having extraordinary jet velocity, devices of fragmentation producing high lethality, and destruction ability systems with the capacity for great blast effects. Normally, cyclotrimethylene trinitramine (RDX) and cyclotetramethylene tetranitramine (HMX) are two compounds in varying compositions used for the development of explosives such as 2,4,6-trinitrotoluene (TNT), HMX, and RDX. Scheme 5.2 represents the structure of TNT, HMX, and RDX [7, 8].

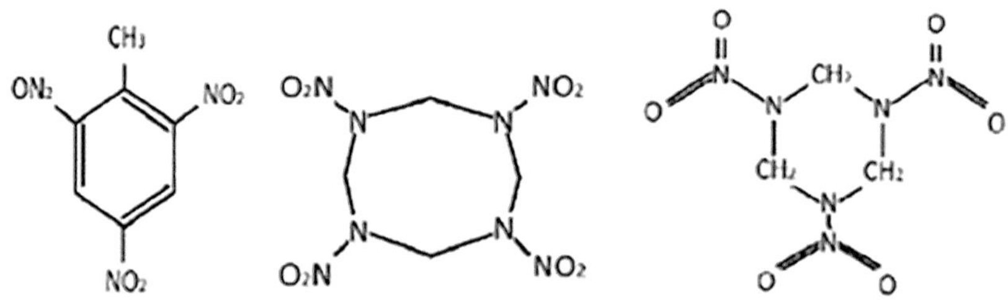

2,4,6 –Trinitrotolune(TNT) **HMX** **RDX**

SCHEME 5.2 Structure of TNT, HMX, and RDX.

5.2 CRITERIA FOR HIGH ENERGY MATERIALS

A good high-energy material possesses the following criteria:

1. These materials release huge amounts of energy at a rapid rate in a self-sustaining chemical reaction, and the energy released is extremely intense and violent.
2. Such materials are often combusted by using the supply of their own oxygen, which may or may not be enough for complete combustion.
3. Hence, for fast and complete combustion with the production of an excessive amount of energy from its constituent atoms and molecules, an oxygen-rich compound or oxidizer has to be added to it.
4. Therefore, HEMs are usually a pure substance or a mixture of substances containing combustible elements or fuel along with

oxidizing elements in a metastable state and also able to react under comparatively minor incitement from a diversity of stimuli, producing heat energy and some gaseous products [9, 10].

5. Although HEMs containing their own oxygen are releasing a comparatively smaller amount of energy, the energy released in the HEMs is much more intensive at a rapid rate, along with some toxic gaseous products.
6. The energy release rate for the given weight and the total energy release during chemical reactions due to combustion distinguish HEMs from other ordinary materials. If the HEMs are appropriately ignited, they can release energy at a comparatively slow rate (deflagration) or at a much quicker rate (detonation); that depends upon the level of the input energy, chemical structure of the material, chemical composition, confinement, and quality of the materials.
7. Hence, those explosive materials that usually have the property of detonation are termed as high explosives, but otherwise, the explosives that are usually deflagrated are called as propellants. The nature, quality, and mechanism of the chemical reactions during combustion of the HEMs depend on the processes of deflagration and detonation, but fundamentally there are no differences. The basic difference between both kinds of materials depends upon the rate and mode of the release of energy and the various phenomena associated with it.

   Example: The process of detonation mainly involves the propagation and formation of the detonation wave, which is normally an intense form of shock wave because of the huge quantities of energy released due to chemical reactions. The shock wave results in a heating effect, which is due to the chemical reaction during the way through the explosive medium at a comparatively much higher pressure of nearly 10 bars with very high velocity. The flame anterior in the deflagrating explosive materials moves at a less linear velocity, normally a few mm/sec, which accelerates to higher velocities with an increase in ambient pressure because of an increase in the degree of confinement.

8. Under some favorable conditions, including the release of a huge quantity of energy and strong confinement influencing initiation,

the deflagration of the propellant can undergo a transition to detonation, but the occurrence of the reverse transition may also be observed. However, the phase due to deflagration is small, which leads to the start of detonation in the high explosive material.

9. The explosives are normally treated as the materials that are capable of experiencing a self-sustained chemical reaction with rapid evolution of heat and release of a huge volume of gaseous products in order to produce the pressure effects.
10. The propellants may be found in all three states: solid, liquid, and gas. The explosives may be a single compound that has the capacity to produce an explosive reaction or a mixture of compounds that may or may not have explosive properties.
11. The molecules of most good explosive compounds must contain adequate amounts of oxygen and nitrogen in their parent carbon skeleton. The role of the nitrogen atoms present in it is to take oxygen atoms from it and initiate the chemical reaction by mixing with hydrogen and carbon atoms at a suitable stress level. The unstable nitrogen-oxygen bonds may sometimes undergo unexpected disruption, initiate the chain reaction, and accelerate detonation or deflagration [11, 12].

5.3 HIGH-ENERGY DENSITY MATERIALS(HEDMS)

This material is a new innovative class of HEMs, and it was recently categorized. There is no clear definition of these categories of materials, but they generally contain HEMs with plasticizer, binder, combustible agent, oxidizer, and high-insensitivity agent. The major necessities of high-energy density materials (HEDMs) for stability and energy are normally about 1.9 g cm^{-3}, a detonation pressure of about 40.0 GPa, a bond dissociation energy of approximately 80–120 kJ mol^{-1} and a detonation velocity of around 9.0 km s^{-1}. These kinds of materials are successfully and efficiently used as explosives, propellants, and pyrotechnics. The innovative advanced HEDMs mainly contain five substances of different categories, such as heterocyclic substances based on N_2 (azine substrates, furazan materials, and azole materials), acyclic compounds, strained materials, caged substances, oxygen clusters, nitrogen cluster molecules, and materials in the excited state. In addition, all the heterocyclic compounds containing nitrogen possess high density, thermal stability, insensitivity, and a high enthalpy, or heat of formation. The compounds containing less hydrocarbon content and high nitrogen molecules facilitate or promote the oxygen balance and increase density. The degraded products of the HEDMs are usually N_2, which has the lowest energy level; therefore, these kinds of materials are environmentally friendly and have attracted the attention of global researchers. The parent chain of the compounds based on azine is mainly the heterocyclic nitrogen-containing aromatic hexagonal ring, and the aromaticity of the compounds is the cause of the greater stability, having less sensitivity toward friction, sparks, and impact and being relatively safer to use. Otherwise, the heterocyclic N-atoms present in the substance replace the CH group of the benzene ring, which causes more heat of combustion, more heat of formation, and more oxygen balance. Presently, more research work has to be carried out on azine compounds, primarily focusing on tetrazines (3,6-dihydrazino-1,2,4,5-tetrazine [DHT] and 3,6-bis(1H-1,2,3,4-tetrazol-5-yl-amino)-1,2,4,5-tetrazine [BTATz]) and amino-substituted diazines or polynitrodiazines (mainly 2,6-diamino-3,5-dinitropyrazine-1-oxide [LLM]-105). The research work on polyazide azo triazines (mainly 4,4′,6,6′-tetra(azido)azo-1,3,5-triazine [TAAT]) and most of the triazines has mainly focused on the medical point of view. High Energy Materials Research Laboratory is the central DRDO laboratory that carries out research for the development, design, and property study of propellants, pyrotechnics, high explosives, energetic polymeric materials, insulating materials, and other HEMs. Scheme 5.3 represents the structure of LLM-105 and TAAT [13, 14].

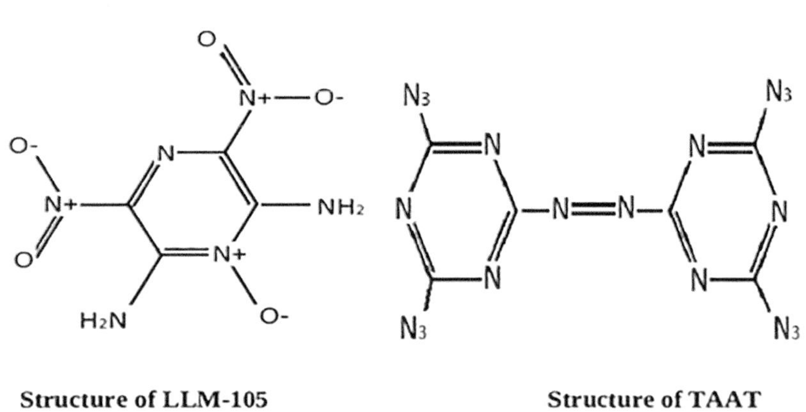

SCHEME 5.3 structure of LLM-105 and TAAT.

The major HEDMs are as follows:

1. LLM-105 has the general molecular formula $C_4H_4N_6O_5$, molecular mass of 216.04, a density $\rho = 1.913$ g cm^{-3} an enthalpy of formation of -12 kJ mol^{-1} and is generally insoluble in most organic solvents except DMSO.
2. TAAT is a polyazido heterocyclic system with many azido groups, and it decomposes exothermically. The nitrogen content of TAAT is very high, up to 79.55%. It has good stability, a high decomposition temperature, and low friction sensitivity. The enthalpy of formation is very high (2171 kJ mol^{-1}). The enthalpy of formation is the highest among all the known polyazide, polynitro, and high-nitrogen compounds.
3. DHT: The compound tetrazine has three isomers, such as 1,2,3,4-tetrazine, commonly known as DHT; 1,2,3,5-tetrazine; 1,2,4,5-tetrazine, or S-tetrazine; or homotetrazine; but chemically, DHT is named as 3,6-dihydrazino-1,2,4,5-tetetrazine, which belongs to the S-tetrazine class.
4. BTATz: From the molecular structure of this material, it was found that this compound contains two heterocyclic high-nitrogen systems, namely tetrazole and tetrazine. It is the compound with the highest nitrogen content among the 3,3′-Azobis (6-amino-1,2,4,5-tetrazine): The compound DAAT is considered a distinctive representative of a high-nitrogen energetic compound called azotetrazine, having a nitrogen content up to 76.36%, low sensitivity, and very good thermal stability. Scheme 5.4 represents the structure of DHT, DAAT, BTATz, and BTT [15, 16].

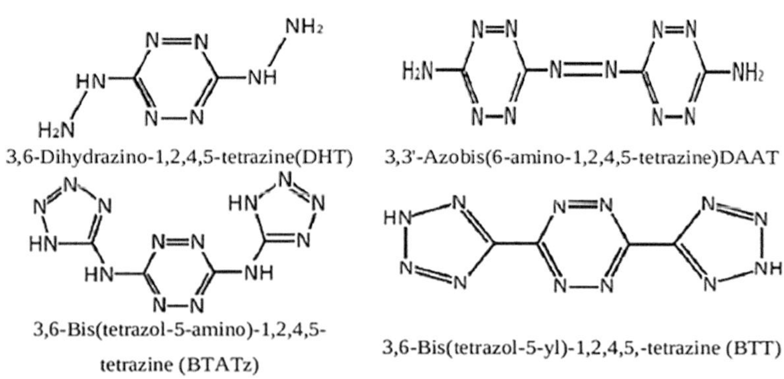

SCHEME-5.4 STRUCTURE of DHT, DAAT, BTATz, and BTT.

5.4 CLASSIFICATION OF HEMS

The HEMs are broadly classified on the basis of the following:

1. Chemical structure of the materials (2) function (3) properties. Based on these properties, the HEMs are generally classified into the following categories:
    - Explosives
    - Propellants
    - Pyrotechnics [17].

5.4.1 EXPLOSIVES

Explosives may be defined in many ways, as follows:

1. Explosive is normally a chemical compound that, on proper triggering, can release a huge quantity of heat energy under pressure due to a vigorous exothermic decomposition reaction in a self-sustaining pathway. The temperature released is usually between 3000 and 5000°C, and the gases released usually expand 12,000–15,000 times in comparison to the volume of the original materials. The whole process takes place within a few microseconds, accompanied by shock and high-intensity noise.
2. Explosives may be defined as a single chemical compound or a combination of chemical compounds that, when subjected to percussion, heat, catalysis, and detonation, undergo rapid degradation associated with the generation of a huge amount of energy along with an unexpected volume of gaseous components greater than the initial quantity of the explosives [18].
3. Explosive can be defined as a chemical material or combination of many chemicals in a device that can produce and release a potential amount of energy on an unexpected outpouring of the gases, thus releasing a much higher pressure on its surroundings.

Recently, scientists and researchers have been searching globally for explosive materials with improved properties such as stability, safety, cost-effectiveness, reliability, and environmental friendliness, and the explosives are generally categorized as follows:

1. Thermally stable or heat-resistant explosives.
2. Explosives having high density and high velocity of detonation (VOD) or high-performance explosives
3. Melt-castable explosives
4. Active binders for the propellant and explosive designs
5. Insensitive high explosives
6. The materials of high energy are manufactured by using N_2O_5

But on the basis of the chemical structure of the explosive materials, they are classified into the following categories [19, 20].

5.4.1.1 NITRO EXPLOSIVES

Normally, explosive of these groups holds the niro group (–NO), which is bonded to the parent carbon chain of the molecule and produces C-NO covalent linkages. Scheme 5.5 represents the structure of pentaerythritol tetranitrate (PETN).

Examples: trinitrobenzene, picric acid, and trinitrotoluene, etc.

The most widely accepted and stable niro explosives include

- TNT
- PETN
- RDX [21].

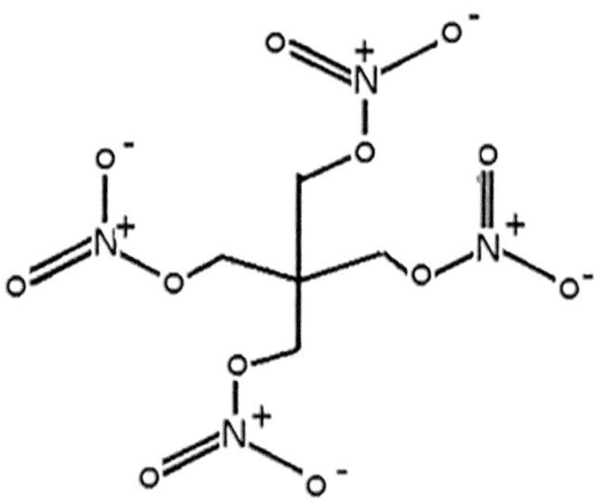

SCHEME 5.5 Structure of PETN.

5.4.1.2 NITRATE ESTERS

In this category of expolosive nitric ester (–ONO) groups, they are bonded with the parent carbon chain of the molecule, resulting in the formation of C–ONO bonds.

Examples: NG, NC, diethylene glycol dinitrate (DEGDN), PETN, polyvinyl nitrate etc. Scheme 5.6 represents the structure of DEGDN and trimethylolethane trinitrate (TMETN).

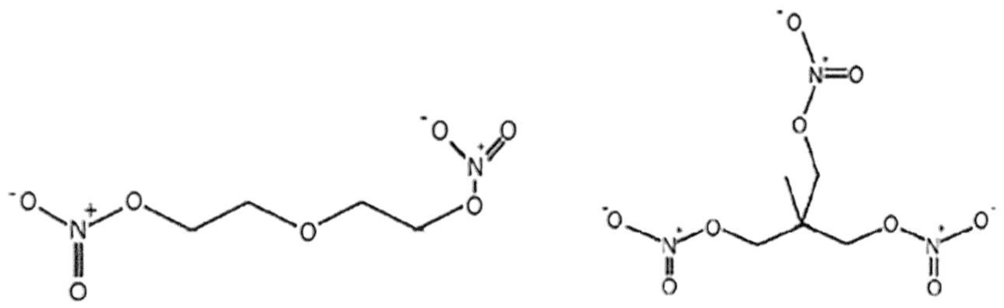

Structure of DEGDN **Structure of TMETN**

SCHEME 5.6 Structure of DEGDN and TMETN.

NG and glycerol trinitrate are the active plasticizers suitably used as explosives on a commercial scale. The other nitrate esters today used as explosives include TMETN, triethyleneglycol dinitrate, ethyleneglycol dinitrate, and butanetriol trinitrate. All these explosives are similar in structure to NG and therefore exhibit a preference almost identical to NG, exclusive of any handling hazards; however, the basic demerit of nitrate esters is their exclusive tendency to autocatalytic decay [22, 23].

5.4.1.3 NITRAMINES

The explosive nitramines contain N–NO groups, which are bonded covalently with the parent carbon skeleton chain and are treated as organic derivatives of the inorganic material nitramine.

Examples: octogen (HMX), cyclonite (RDX), etc.

Some of the materials, which by themselves do not show the characteristics of explosives, are, however, the cause of oxygen carriers in many explosive mixtures, so they may be treated as propellants or pyrotechnics.

Examples: ammonium nitrate nitrate (NH_4NO_3), potassium nitrate (KNO_3), sodium nitrate ($NaNO_3$), ammonium perchlorate (NH_4ClO_4), potassium chlorate ($KClO_3$), etc. [24, 25]

5.4.1.4 AZIDO

The explosives in these categories are usually energetic plasticizers associated with azido-acetate esters. These kinds of explosives exhibit a lower T_g value, excellent thermal stability, and high compatibility.

Examples: trimethylolnitromethanetris (azidoacetate) (TMNTAA), ethyleneglycolbis (azidoacetate) (EGBAA), pentaerythritol tetrakis (azidoacetate) (PETKAA), diethyleneglycol Bis (azidoacetate) (DEGBAA), etc.

The common demerits of these kinds of explosives are a lower oxygen balance, an increase in the heat of formation, and a high density. The existence of heat in formation is due to the creation of a number of carbonyl groups, which partly balance the azido groups. Scheme 5.7 represents the structure of DEGBAA, EGBAA, PETKAA, and TMNTAA [26].

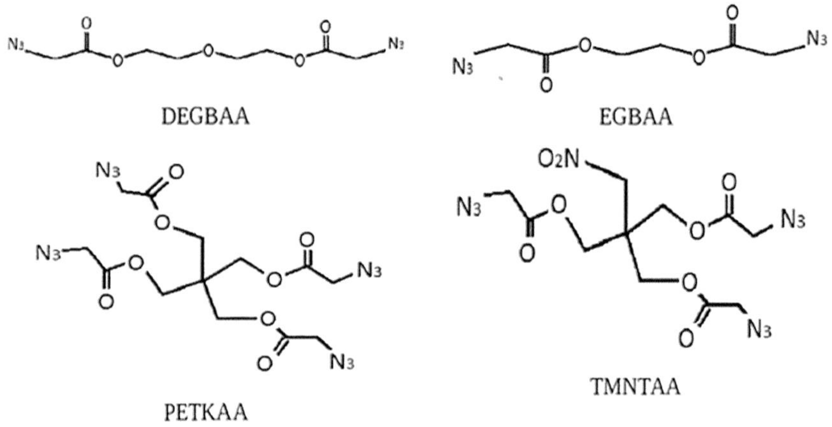

SCHEME 5.7 Structure of DEGBAA, EGBAA, PETKAA, TMNTAA.

Again explosives are classified on the basis of their intensity of function, which are as follows:

5.4.1.5 DETONATING EXPLOSIVES OR HIGH EXPLOSIVES

1. The chemical substances released either on decomposition or combustion release enormous amounts of energy with huge amounts of numerous toxic gaseous componets which are termed as explosives. Usually explosives are broadly classified into two categories, such as high explosives and low explosives.

2. The explosives that react at a very rapid rate, creating a large amount of heat energy, and some gaseous compounds are termed as high explosives. During the process of explosion, supersonic shock waves are produced, which can pass through the molecules of the explosive and facilitate faster combustion. These are usually composed of a single specific substance or a mixture of appropriate substances. It possesses the properties of groups or atoms of the reacting molecule, leading to an explosion with the generation of large quantities of energy.
3. The most common high explosives used commercially include HMX, NG, RDX, TNX, HNS, TNT, PETN, and TATB.
4. Explosives such as TNT and dynamite are detonating in nature and can undergo tremendously quick degradation with unexpected high pressure; however, deflagrating explosives, although they burn at a faster rate, rather generate low pressures. Therefore, high explosives are usually treated as the kind of explosive that possesses the capability of detonating at a rate similar to supersonic velocity. The rate of detonation of high explosives is normally 3–9 km/s.

   Example: The detonation rate of TNT is normally 5.8 km/s, and the detonating cord is about 6.7 km/s.

5. The high explosives are specifically used in mining, demolition of rocks, and some particular military applications. According to their sensitivity, high explosives are broadly classified into two kinds [27, 28]:

   1. Primary high explosives (initiating explosives)
      Although these categories of explosives are highly active, they are relatively less powerful.

      Examples: lead styphnate (LS), mercury fulminate (MF), lead azide (LA), diazodinitrophenol, etc.

      The explosive materials in this category are extremely sensitive to friction, mechanical shock, and heat. It is highly combustible and very quickly detonates, making it highly effective for military applications and blasting. Such types of explosives are used in the devices as primers or detonators and are commonly called as primary explosives. The most commonly used primary high explosives are LA, MF, and LS. These kinds of explosives exhibit highly hazardous effects not only to humans but also to the whole environment. However, no suitable alternative explosive as compared to the stability, performance, and sensitivity of these kinds of explosives has been developed or synthesized till now [29].

2. Base explosives or secondary high explosives

   The secondary explosives are less sensitive but more stable as compared to the primary explosives. Such kinds of explosives are relatively insensitive to friction, shock, and heat. Although such varieties of explosives are not easily initiated, they exhibit the capability of powerful detonation.

   Examples: octogen (HMX), tetryl, RDX, TNT, and hexogen.

This kind of explosive is comparatively more successful for detonation because it is always associated with high sources of initiation energy, which can be transmitted directly via boosters of secondary explosives or from the primaries. There are several suitable mechanisms suggested for the initiation and function of primers and detonators, but all these mechanisms depend upon the sensitivity, stability, and performance of the secondary explosives. From time to time, these kinds of explosives are modified by continuous development in the process of initiation, level of performance, quantity, chemical composition, and some other essential mechanisms. The most commonly used primaries include LS, barium nitrate, LA, antimony sulfide, a mixture of diazodinitrophenol, and tetracene, which are highly poisonous, stable, sensitive, and extremely hazardous to our ecosystem. In spite of so many demerits, these explosive materials are still used as the primary choice because of the lack of alternatives. Hence, extensive research must be required to develop green explosives or eco-friendly explosives to protect our environment from this hazardous effect [30, 31].

5.4.1.6 BLASTING AGENTS OR TERTIARY HIGH EXPLOSIVES

These classes of explosives are extensively intensive to shock and detonate at a steady rate and are preferred as the midway of the explosive booster linked with secondary explosives like ammonium nitrate, fuel oil mixture, and slurry explosives. These explosives are mostly used extensively for mining and construction activities.

Examples: compressed forms of acetylene, dynamite, NC, picric acid, cyanogen, trinitrotoluene, nitroglycerin, perchlorate ammonium nitrate, wood pulp mixed with liquid oxygen, nitroguanidin, and nitrobenzene mixed with fuming nitric acid [32, 33].

5.4.1.6 GREEN EXPLOSIVE

1. These kinds of explosives are absolutely new and improved categories of explosive substances that are normally eco-friendly toward the environment, nonhazardous to human health, and maintain sustainability toward the ecosystem.
2. Today, scientists in the US designed and developed a new class of explosive materials that are fairly eco-friendly, sustainable for the environment, and can be used as a proper substitute for toxic lead-based primary explosives. These green explosives can be used for detonating purposes in all the blasting caps of ballistic missiles. The technique of manufacturing these kinds of explosives follows suitable green technology, which is comparatively more safe.
3. Primary explosives are generally less powerful in their performance, more sensitive, stable, and effective in the case of powerful explosions. But primary explosive does not fully satisfy its performance because it is less powerful; therefore, mercury fulminate and lead-based chemical compounds are added to it, which are highly hazardous to the people associated with this work. It also has a high adverse impact on the environment. It was reported that 10 times more lead was found in the blood of the people working in the gun range of the FBI associated with this kind of explosive as compared with the permissible range as per government safety regulations.
4. Till now, the scientists have struggled to develop a proper substitute exposure material that is environmentally friendly and exhibits good stability and sensitivity.
5. Later, a new primary explosive was developed by using nitrotetrazole. Although, in combination with the substances based on perchlorate, it serves as an excellent primary explosive,

substitution of the charge groups exchanging in the molecules of the substance changed its overall sensitivity to ignition. However, its sensitivity decreases if it is in contact with an aqueous medium. This is the main demerit of this category of explosive as compared to the other common primary explosives [34, 35].

6. The first eco-friendly, green primary explosive based on the toxic heavy metal Pb was synthesized by researchers at Los Alamos and North Carolina. Thomas J. Meyer developed four types of good, eco-friendly primary explosives. These explosives fulfil the criteria needed to substitute for the explosives associated with lead. The explosive materials of these kinds normally contain complex di-anions of metals and cations of environmental benignity. These explosive materials can be diffused by storing them in water without releasing any toxic by-product materials when they are blown up.
7. 5-Nitrotetrazolato-N_2-ferrate is treated as a basic green primary explosive. Today, tetrazole oxide 1*H*,1′*H*-[5,5′-bitetrazole]-1,1′-diol (H_2BTO) in combination with the energetic coordination polymers (ECPs) with the help of metal cations like Mn(II), Zn(II), Ni(II), Cu(II), and Co(II) results in five various kinds of interesting green primary explosives devoid of any organic toxins or toxic metal ions.
8. Therefore, it was confirmed that green ECPs are normally a new, superior kind of primary explosive that does not contain any concentration of lead.

   Example: The thermal decomposition of such kinds of green explosive materials is >200°C, and on the basis of the thermal resistance of these substances, it was identified that explosives of these categories exhibit long-term stability at temperatures less than 100°C. These materials detonated in a controlled way, making them safer than other explosives. ECPs possess the capability of a quick deflagration-to-detonation transition. Some other environmentally friendly primary explosives are cat[FeII$(NT)_3(H2O)_3$], cat4[FeII$(NT)_6$], cat2[FeII$(NT)_4(H2O)_2$], and cat3[FeII$(NT)_5(H_2O)$].
9. The environmentally friendly green explosives are usually free from lead and other toxic materials and exhibit the property of

thermal instability. Even though a huge number of energetic material combinations are being designed and developed, such as coordination complexes, organic salts, zwitterions, specific organic materials, specific inorganic salts, and metastable interstitial composites, none of these fulfil the challenges and needs of green primary aspects such as:

1. Never respond and be stable toward moisture and light.
2. Although at the beginning it is sensitive, later on it is not appreciably sensitive to control, handling, and transport.
3. These kinds of explosives are thermally stable up to a temperature of 200°C.
4. These explosives exhibit excellent thermal stability for a longer period of time.
5. Does not contain any toxic, unfavorable heavy metals like Pb, Ba, Sb, Hg, and Ag.
6. Does not contain any perchlorate substances [36, 37].

5.4.2 PROPELLANTS (OR LOW EXPLOSIVES)

1. Propellant is commonly called a low explosive, which is normally the combination of an oxidant and a combustible material and has the capacity of decomposition or deflagration very rapidly; however, it burns comparatively more slowly than a proper high explosive. Low explosives are more commonly used as propellants.
2. These materials are combustible in nature, holding the required oxygen within them for combustion. They burn steadily without undergoing explosions and produce a huge volume of gases, which are the cause of the explosion. In the usual situation, low explosive materials undergo deflagration at a rate in the range of only a few centimeters per second to almost 400 m s^{-1}. But it is feasible to deflagrate rapidly, creating a consequence comparable to detonation. This class of explosives is mainly ignited in a restricted space area, and the rate of transmitting energy differs very widely depending upon the nature of the materials [38].

Examples: black powder, flash powder, smokeless powder, NC powder, cordites, liquid oxygen, polybutadiene, liquid hydrogen, etc.

On ignition, these categories of explosive materials only burn in a steady state, do not undergo an explosion, or never detonate, but produce huge volumes of gases with a moderate pressure of up to 3000 kg/cm.

1. Propellants mainly consist of oxidizers and fuel materials. One important example of a low explosive is black powder, which is normally used for fireworks and burns very rapidly. The black powder is generally kept at the interface of the paper tube or cardboard, and once it ignites, the gases produce rust and create a noisy environment. The lower explosive materials or propellants are engineered in such a manner that they react judiciously at a controlled rate and generate the effect of pressure at a sustained rate for a longer period relative to high explosive materials.
2. Like high explosives, the propellants use several various kinds of materials in the process of igniting. Propellants are formulated either in the solid state or in the liquid state. Solid propellants are most often used in guns, smaller rockets, and cannons, whereas liquid propellants are mainly used in high-performance missile systems and some other specific applications. Again, solid lubricants can be categorized on the basis of their chemical composition, and each class has its own exclusive properties and can be applied in definite applications. All solid propellants contain some additives, and these additives are comparatively more toxic than the original components of the propellant and pose an occupational hazard to humans and the environment.
3. Solid propellants have advantages over liquid propellants regardless of class and composition, which include their compactness, ease of storage, safety, tolerance at extreme temperatures, and ease of handling. But in comparison to that, the use of liquid propellants is limiting their use in high-performance missile systems [39, 40].

5.4.2.1 TYPES OF PROPELLANTS (LOW EXPLOSIVES)

On the basis of the use of propellants for the thrust of a rocket, gun projectile, or vehicle, they are classified into the following two groups:

1. Gun propellants

The composition of the gun propellants is generally a mixture of two or more energetic substances in addition to various categories of additives, which work specifically and play a definite role. The gun propellant is used in the cartridge of a bullet and can be ignited by a primer cap or suitable igniter and operated in the gun chamber, creating a pressure up to 450 MPa with a very short combustion time in the order of a few milliseconds. Now it is a challenge for researchers worldwide to develop the performance of the propellants. The propellants having low flame temperatures and a soft launching capacity show good performance. On the basis of the composition of the gun propellants, they may be subdivided into the following three categories [41, 42].

a. Single-base propellants

Presently, NC is considered the only single energetic ingredient material. In addition to that, a definite amount of plasticizer or gelatinizer is added as a stabilizer in the final formulation to control the chemical stability and enhance the properties of the propellants. Its detonation velocity is in the range of 7300 m/s (23,950 ft/s). The ether-alcohol and colloidal solution of NC is the only explosive propellant ingredient [43].

b. Double-base propellants

This contains NC and NG as the major and energetic ingredients; some liquid organic nitrate explosives may be added to improve the performance. In addition, small quantities of plasticizer, stabilizer, etc. are present. These are more energetic than single-base propellants. Stabilizers and some other suitable additive materials are added depending on the use of the propellants. The role of NG is to increase energy output and reduce smoke formation. These propellants are mainly used in mortars, small arms, cannons, and rockets [44].

c. Triple-base propellants

These kinds of propellants contain three constituents, of which nitroguanidine is the major constituent and the other two are NG and NC. In addition to the three constituents, some additive materials and stabilizers are added depending on the use.

Some nitramine compounds, such as HMX and RDX, are also being added as energetic ingredients to increase the performance of some modern and advanced gun propellants, which are used in ammunition for tank guns. Recently, liquid propellants have been developed that are more successfully used in gun propulsion. The most talented liquid gun propellant is a water-soluble mono-propellant containing alkyl ajnine nitrate, hydroxyl ammonium nitrate, and water [45].

d. Composite propellants

1. This category of propellant does not contain NG, NC, nitroquanidine, or the compounds of organic nitrate. These composites normally consist of a fuel like metallic aluminium, binder substances like synthetic rubber, and an oxidizer like ammonium perchlorate. These are an advanced category of propellants and are mainly used in large rocket motors [46].
2. Rocket propulsion is a device or system in which the propellant system is usually ignited and burns in the combustion chamber, releasing hot gases or working fluids that are accelerated and expanded through a divergent–convergent designed nozzle and produce a thrust or force due to a combustible reaction in the opposite direction.
3. A rocket engine is usually operated by using the principle of Newton's third law of motion. The physical states of the propellants are different before and after combustion, such as liquid, solid, or hybrid state, that is, a combination of both liquid and solid. The pressure on the operating condition inside the combustion chamber usually ranges from 2 to 20 MPa; however, the time of combustion varies from one second to several minutes, which depends upon the operating condition and application.
4. There are many chemical substances that are used as components of liquid rocket propellants, but their use is limited to military propellants. Liquid rocket propellants are normally used exclusively in military devices, such as nitrogen tetroxide, hydrazine, and boranes. Although hydrazine is a widely used chemical substance in the liquid propellant industry, it has a great impact on both human and animal health systems.

The impact on human health is restricted to the irritation of our skin and hepatic and mucosal disorders, but its impact on animal health security is more severe.

Example: Hepatomas are developed in mice if they take hydrazine; rats exposed to the vapors of hydrazine suffer nasal tumors; and hamsters also develop lung adenomas.

5. The periodical assessment of the erythrocyte indices in the least medical surveillance indicates the different diseases, such as kidney, hypoglycemia, liver disease, hemorrhagic diathesis, and allergies if exposed to phenylhydrazine and isoniazid. Nitrogen tetroxide is a prominent chemical widely used in the space program and is the cause of a potential health hazard. The vapors of it may be the cause of instant or delayed blistering and swelling of the adnexa oculi, as well as severe burns of the dermis. If we take nitrogen tetroxide in the process of respiration, then it reacts with the moisture present in the lung tissue and forms HNO_3, which is the cause of delayed pulmonary edema. Boron hydride and its related derivatives, commonly called boranes, are used as fuel in rocketry, as vulcanizers for rubber, corrosion inhibitors, and some of the components.
6. Boranes have a significant toxic effect. It is more toxic to the cardiovascular system, respiratory system, skin, CNS, kidneys, and liver. Carboranes—boranes are another compound that contains carbon, hydrogen, and boron and is developed especially for use in solid-fuel systems. It causes irritation to the skin and is comparatively less toxic [47, 48].

Rocket propellants can be suitably divided into the following classes:

5.4.2.2 SOLID PROPELLANTS

1. Propellants in this category are normally prepared from diluted or low-level explosives. These are only deflagrated, not detonated. The solid propellants mainly consist of an oxidizer, a fuel, and a binder material and are mixed in the form of grains. Solid propellants

are usually easier to use, may be put in storage for a prolonged period, and are basically suitable for use by the military in missiles.

2. Since these categories of propellants have comparatively low performance and do not impact sufficient thrust, they cannot be suitable for space rockets. Propellants of these kinds are comparatively more stable and appropriately stored for war during fighting, whereas liquid propellants are more energetic, cryogenic, and regularly used in satellite launch vehicles. Solid propellants are more significant and frequently used in cannons, guns, and smaller rockets, whereas liquid propellants are mostly used in high-performance missile systems and also in certain other specific applications.
3. The solid propellants can be categorized on the basis of their composition, and each class of propellant has its own exceptional properties, which are the reason for their usefulness in different specific applications. All solid propellants contain some additive materials, and these materials are generally more toxic and harmful to human health and the environment than the original materials. Solid propellants are safe, easy to manufacture, less expensive, easily storable for a longer period, can function and maintain stability at extremely elevated temperatures, and are easy to handle.
4. The solid propellant consisting of ammonium nitrate as an oxidizer and poly(GLYN) as a binder, along with small quantities of aluminium and boron, provides optimum performance even without plasticizers in large launching vehicles, space motors, gas generators, etc.
5. The solid propellant consists of ammonium nitrate oxidizer, poly(GLYN) as a binder, and Al or Mg fuel elements, which show comparatively high performance.
6. Similarly, the solid propellant, consisting of ammonium perchlorate as an oxidizer, poly(GLYN) binder, and beryllium hydride or beryllium as fuel material without plasticizers, also functions at optimum performance even at low solid levels and is suitable for space-launching vehicles [49, 50].

Due to the mixture of different materials, it may be again subdivided into various categories.

1. Double base propellants or homogeneous propellants
   These types of solid propellants contain a mixture of NG and NC, with some additive materials depending on the applications.
2. Composite propellants or heterogeneous propellants

   Propellants in these categories contain an oxidizer, which is uniformly dispersed within a binder matrix, and some additive materials. The oxidizer used is a finely powdered form of inorganic oxy-salt such as ammonium perchlorate, ammonium nitrate, sodium nitrate, etc., among which ammonium perchlorate is the most widely used substance. The binder material currently used is normally organic polymers such as polybutadiene, polyurethane, polysulfides, etc. In a few cases, metal fuel, such as a finely powdered form of aluminium, may be added to enhance the performance of the propellants [51].
3. Composite modified double base propellants

   These types of propellants normally contain a fine powder form of ammonium perchlorate and aluminium particles, which are dispersed in the matrix of NG and NC.

Merits of solid rocket propellants

1. The solid propellants used in the rockets are relatively easier to handle and store than the liquid propellants.
2. The propellant density is higher, which makes it compact in size.
3. Because of their low cost and simplicity, the solid propellants used in rockets are considered ideal propellants for military applications.
4. The simplicity of solid propellant rockets makes them a suitable choice if more thrust is required at a low cost [52].

Demerits of solid rocket propellants

1. The use of these propellants as rocket fuel usually exhibits less impulse in comparison to the liquid fuel used in rockets.
2. Due to the high density of propellants, the high mass ratio is observed with the strength-to-weight ratio of filament-wound motor casting and decreases the performance.
3. The solid rocket propellants may not be throttling at the stipulated rate.

4. Solid propellants used in rockets are capable of removing the reverse thrust or combustion within the accumulating warhead separation or range of control.
5. The casting of larger quantities of propellant required repeatability and consistency, which can be assured by computer control.
6. The rockets with solid fuel are normally intolerable to voids and cracks, and the fault was identified by X-ray scanning after post-processing [53, 54].

5.4.2.3 LIQUID PROPELLANTS

1. Liquid rocket propellants, or the propellant used in the engines of rockets, are liquid forms of propellants. Liquid propellants are generally more preferred than solid propellants because of their high energy and smaller tanks. When fuel and oxidizers are both mixed in the combustion system, burning of the fuel occurs. The Liquid propellant used in rockets may be monopropellant, bipropellant, or tripropellant, but the bipropellant is mainly used as the liquid oxidizer and liquid fuel for the rocket engine.
2. The liquid propellants are more necessary due to their relatively greater density, which permits the propellant tank volume to be relatively smaller and can be made possible by using centrifugal turbopumps of lightweight to deliver the liquid propellants from the tanks into the combustion chamber; that is, the propellants may be stored at relatively low pressure. Since the propellant tank is lighter, it is responsible for the high mass ratio of the rocket.
3. The inert gas that is stored in the tank under high pressure may occasionally use pumps in small and simpler engines to force the propellants inside the combustion chamber. Although these small engines have a lower mass ratio, they are normally more consistent and reliable, so they can be used broadly in satellite systems for maintenance and control of orbit.
4. The liquid rocket propellants may be monopropellant, which uses a single category of propellant; bipropellant, which uses propellants of two different categories; or comparatively more unexpected, tripropellant, which uses propellants of three different categories. But practically, rockets normally use bipropellant liquids as fuel and oxidizers in liquid states, such as liquid oxygen, liquid

hydrogen, or hydrocarbon fuel. The rocket engine is a cryogenic engine in which the oxidizers, such as liquid oxygen and hydrogen, and the fuel are in a gaseous state [55, 56].

1. Monopropellants

   In monopropellant fuel, a single substance serves as both fuel and oxidizer. These compounds are either well-ordered, such as hydrazine, isopropyl nitrate, or ethylene oxide, or in the form of a homogeneous solution, such as nitromethane-methanol, commonly called myrol. Their applications in rockets are partial and mostly auxiliary [57].

2. Bipropellants

   a. These categories of propellants are combinations of fuel and oxidizer, where both substances are stored separately in two different storage tanks and inserted into the combustion compartment during the time of the rocket's launch. The oxidizers used are mainly N_2O_4, H_2O_2, red-fuming HNO_3 (RFNA), and dinitrogen pentafluoride. There are a variety of substances used as fuel for bipropellant rockets, such as hydrazine, orthotoluidine, unsymmetrical dimethylhydrazine, diethylene triamine, triethyl amine, mixtures of xylidines and triethyl amine, RP kerosine, liquid fluorine, and liquid oxygen. The latter two are cryogenic, whereas all the others are storable.

   b. The pairs of many fuel oxidizers are hypergolic or self-igniting. For applications in space launching, mostly nonstorable or cryogenic liquid propellant combinations are preferred because of their high performance level, whereas for military use, storable oxidizer-fuel combinations are normally preferred because they are easy to handle [58].

3. Tripropellants

   a. A tripropellant rocket is basically a rocket whose engine can be operated by using liquid propellants that combine three different categories of substances. The tripropellant rocket engine system is designed and developed in such a manner that it provides high specific impulse and has been considered for the design of a single stage for orbit.

b. The tripropellant rockets are divided into two different types. One type of engine combines all three separate streams of propellant and burns all three simultaneously. The second category is one that is the combination of two different types of fuels and one oxidizer; the two fuels are burned in sequence during the flight.
c. By using fluorine and lithium without hydrogen, the highest specific impulse observed was in 458 seconds, but in comparison, if fluorine and lithium burn stoichiometrically along with hydrogen, then 30% of the mass flow was measured [59].

5.4.2.4 HYBRID PROPEL1ANTS

This is a modern and improved class of propellant, which is the combination of liquid oxidizer with solid fuel or solid oxidizer with liquid fuel. The former is called a standard hybrid, whereas the latter is called a reverse hybrid.

Example: The propellant of standard hybrids is aniline formaldehyde/RFNA, which may be capable of hypergolic or spontaneous ignition.

The standard dual-vertical-flow hybrid class of rocket engine design can also use N_2O/HTPB, H_2O_2/HTPB, N_2O/HDPE, or H_2O_2/HDPE combinations of propellants [60].

Advantages of liquid rocket propellants

1. Rockets using liquid fuels are normally more impulsive in comparison to those operating on solid fuel.
2. These fuels are able to choke, restart, and shut down.
3. The liquid propellant vehicles generally use turbopumps, and the propellant containers are usually at extremely low force in the combustion chamber. This is the reason for the orbital launching vehicle application, preferably by using liquid propellants.
4. The major advantage of liquid propellants is mainly because of the oxidizer. A number of different kinds of liquid oxidizers, such as liquid oxygen, H_2O_2, N_2O_4 are available, which have been found to have excellent individual impulse in comparison to ammonium perchlorate (NH_4ClO_4), which is normally employed in most of the solid-fueled rockets.

5. The liquid propellants are comparatively less expensive than the solid propellants. Certain propellants can absorb oxygen and nitrogen from the uppermost layer of the troposphere, transfer these up to the lower orbit of the earth, and be used as depots of propellants significantly at a reduced rate [61].

Disadvantages of solid rocket propellants

1. These propellants are normally the least abstemious and the most difficult to handle and store because of their lower reactivity with conventional materials.
2. Numerous categories of exotic oxidizers can be projected, although these are energetic but unstable and toxic in nature.
3. The liquid-fueled rockets are potentially tolerable, requiring seals and a combustion chamber that is thermally stressed, which is therefore the result of the increasing cost of the rocket [62, 63].

5.4.2.5 GASEOUS PROPELLANTS

The gaseous propellants normally contain some special classes of compressed gases, but because of their low density and heavier pressure container, they are rarely used. This class of propellants is sometimes used in lighter engines, but it is practically used with the inert propellants [64].

Merits of gaseous propellants

1. These propellants undergo combustion easily.
2. These propellants are less costly and readily available.
3. The gaseous propellants lack density.
4. It can be stored for a longer period of time.

Demerits of gaseous propellants

1. The pressure produced in the container decreases the lifespan of the aerosol [65].

5.4.3 PYROTECHNICS

1. Pyrotechnics is the phenomenon of using materials that are efficient for undergoing self-sustained and self-contained chemical reactions (exothermic) for the generation of heat, gas, light,

smoke, and sound. Pyrotechnics is not only the production of fireworks but also some other objects, including oxygen candles, safety matches, explosive fasteners and bolts, components of the automotive airbag, and gas pressure blasting in the mining sector, demolition, and quarrying.

2. When these substances are ignited, they produce fire, smoke, light, and/or sound, as well as flammable effects. These substances are usually a combination of finely powdered forms of combustible materials, including oxidants and binders [66].

   Examples: gunpowder, magnesium powder, varnish/$NaNO_3$, Zr/$KClO_4$, some seed oil, etc.

3. These classes of materials have a comparatively slow rate of burning and are nonexplosive powder forms of substances, which include alloys, metals, and mixtures of hydrocarbons. In this chapter, we are discussing the pyrotechnic substances that are used in propellants. However, pyrotechnic materials are extensively used in various military devices, including flares, relays, signals, fuses, delays, metal-parts plants, small-arms plants, and shell loading, assembly, and packing (LAP) plants.
4. This material is not only used in ammunition plants but also in the manufacturing of weaponry, testing, storage in the arsenals, testing munitions at proper ground areas, and maintaining the weaponry demilitarized at the workshops. Workers exposed to such substances suffer toxic and occupational hazards.
5. Although pyrotechnics are best applied to fireworks, they are still used in different military and civilian systems. On the basis of this chapter, pyrotechnic material may be defined as the combination of an oxidant with a solid fuel, which is accomplished by self-sustaining combustion even without air.
6. Pyrotechnics, in addition to propellants and explosives, form high-energy materials that are classified between the three different kinds of materials as related to the nature and condition of the burning process. Pyrotechnics are generally processed due to deflagration, that is, the propagation from layer to layer in contrast to explosive substances, where combustion is caused by detonation and the creation of shock waves.

7. The traditional types of pyrotechnics are generally a combination of oxidants and fuel. The fuels are mainly metals such as Al, Mg, and Fe, and nonmetals such as Si, C, and S, as well as certain organic compounds. Oxidants used included peroxides, oxides, and different salts. Various types of additives are added to enhance specific properties. Now a more advanced type of pyrotechnics has developed, which is made of bimetallic alloys with resin-bonded or polymeric materials.
8. Today, gasless pyrotechnics are developed and designed with reproducible burn times. It is a great evolution that less than 10 cm^3 of gas is evolved during the burning of 1 g of the pyrotechnic mixture.
9. A technique for the production of pyrotechnics was designed by using the methodology of evaporation of Mg at a comparatively high rate with the help of PTFE. Here, Mg serves as fuel, and the F2 present in the PTFE acts as an oxidizer. During ignition, the PTFE melts and releases fluorine, which immediately reacts with Mg, producing a huge amount of heat energy, which would help sustain the process. The byproducts obtained in this process are carbon and magnesium fluoride [67, 68].

5.5 PROPERTIES OF EXPLOSIVES

The explosives are studied based on their properties like detonation velocity, detonation pressure, water resistance, density, and fume type.

1. Density: The density of the explosive is measured based on the specific gravity, which is the ratio between the density of the explosive and the density of the water at normal atmospheric conditions. For commercial types of explosives, the specific gravity ranges from 0.6 to 1.7 g/cc, whereas for free-administration explosives, the density is normally quantified as the pounds of explosives per foot of the charge length in a specified size of borehole. For tough blasting where fine fragmentation is needed, high-density explosive is preferred, whereas for easily fragmented rocks where fine fragmentation is not required, low-density explosive is the preferred one.
2. Borehole and detonation pressure: The detonation pressure is normally a function of the detonation velocity and explosive density. It was experimentally found that the detonation pressure

is comparatively more dependent on the VOD than the specific gravity. The detonation pressure should be higher if the rock is denser and the blasting is tough, whereas in the case of soft rocks, a low detonation pressure is sufficient. The detonation pressures of the explosives generally range from 10 to over 140 kilobars.

3. Resistance to water: The measure of the stability and sensitivity of an explosive depends upon its ability to withstand water. Sensitivity is the property of detonating an explosive. In the case of dry work, the property of water resistance has no meaning, but if the water is present in the borehole and the time in between firing and loading is very small, then the water resistance explosive is rated as a good explosive, but otherwise, if the water is penetrating into the borehole, then, in that case, the water resistance explosive is considered a good explosive. The explosives of high density have good resistance to water, whereas the explosives of low density, or blasting agents, are less resistant to water [69, 70].
4. Fume class of explosive: The commercial explosive, on detonation, usually produces CO_2, N_2, and water vapor along with some unwanted toxic gases such as CO, NO_x. All these gases are known as fumes, and the explosives of the fume class specify the quantity and properties of the unwanted gases released at the time of detonation. The explosives that emit the least amount of fumes are considered to be of higher quality. When working in an open space, fume release is not so important, but when working in a closed space, the fume release rating for an explosive is very important. The CO gas slowly affects our central nervous system and brain, whereas the NO_x immediately forms HNO_3 in the lungs and may be fatal to us [71].
5. Shelf life: Most of the explosive materials have a minimum shelf life of 1 year, but otherwise 2, 3, or even 4 years of shelf life are rated as good explosives. One year, or maximum after storage of 2 years, is the shelf life measured [72].

Permitted or permissible explosives:

An explosive is said to be permissible if it is approved by the British Ministry of Fuel and Power or the US Bureau of Mines. When the explosives are exploded or detonated, they produce flames of different intensities, volumes, temperatures, and durations.

Example: Black powder creates a low-intensity flame and burns for a longer time, whereas dynamite produces small, lasting flames, but the intensity of the flame is much higher.

The permissible explosives are mainly designed and manufactured for low-volume flames, low temperatures, and short durations. These explosives are prepared by the addition of some specific salts to the explosive materials for proper cooling and quenching the flame to inhibit the ignition of dust or gas in the restricted space of the mines. The permissive explosive materials are usually some modified kinds of emulsions, ammonia dynamites, and water gels [73, 74].

5.6 TESTS OF THE EXPLOSIVE MATERIALS

To identify whether a single material or mixture is an explosive material or not, it was tested according to the following standard procedure:

1. Test for sensitivity to friction or impact: A steel ball of about 100 g by weight was dropped into a few milligram samples from a certain specified height. If the material explodes, then it must be sensitive to impact. Then, the test for friction is normally done by taking a sample of the explosive and placing it between two steel plates along with some sand, and then allowing the upper plate to rotate against the lower plate, which is not movable.
2. Thermal stability: A sample of about 20 mg of explosive materials is taken in a bath at a constant temperature. If the material explodes within a period of 10 seconds, then it is suggested that the explosive is sensitive to heat.
3. Velocity of detonation: Normally, the VOD of the explosive materials can be measured by using high-speed photography, direct flame photography, and counter chronographs.
4. Destruction power: This test is more commonly called the lead block test. This test is carried out by taking a sample of about 10 g of explosive material, which is kept in a hole that is bored in the central portion of the lead block. The explosive material inside the hole was sealed with sand, and then the explosive was allowed to detonate. The power of destruction of the explosive material is measured by estimating the volume of the hole within the lead block. The larger volume indicates the high power of the explosive [75, 76].

5.7 MILITARY EXPLOSIVES

Military explosives are explosive compositions and explosives or formulations that are normally used in military weaponry such as bombs, torpedoes, shells, grenades, rocket warheads, and missiles. The weapons carrying secondary explosives are normally more insensitive to some extent and hence safe for control, handling, transportation, and storage. These explosive materials are mainly formulated by means of an explosive carriage containing an initiator and a booster or intermediates. The explosives are treated as good military explosives if they are stable both chemically and physically over a range of humidity and temperatures for a longer period of time. These explosives must be judiciously insensitive to the impact that is skilled by the weaponry shells during the firing of the gun or when they enter the steel armor. These explosives are used during the firing of projectiles within the aerial time bombs without an impulsive explosion. The required raw materials to produce such kinds of explosives must be widely available for manufacturing in bulk during times of war. The explosives for general applications are divided into four different kinds:

1. High explosives or detonating
2. Low explosives or deflagrating
3. Pyrotechnics
4. Commercial or civil explosives [77, 78]

5.8 RECENT DEVELOPMENTS OF HEMS

HEMs are normally complex systems, and their use first depends upon modeling the system and, second, the methods used to solve the problems associated with these materials. The problems based on HEMs are basically designed for nonlinear integer programming, linear programming problems, mixed-integer linear programming (MILP), nonlinear programming, and its variants. A multitude of techniques have been adopted in recent years for HEMs, but the technique selected depends upon the nature of the issue.

Linear programming: This programming is an optimization technique in which the objective function is a linear function and the constraints applied are also linear functions. It is the simplest method to represent mathematical programming. Linear programming is computationally expensive and difficult to model HEMs; therefore, it is not always possible to apply this programming to the system.

Mixed-integer linear programming: MILP usually belongs to the optimization techniques, where the objective function feed is a linear function and is subject to linear constraints; however, it uses continuous, mixed, and discrete variables. These problems permit improved modeling power because it is possible to take binary variables, which help to denote real-world procedures comparatively more efficiently. The formulation and modeling of HEMs are done more effectively by using MILP.

***Nonlinear programming*:** Nonlinear programming is completely distinct from MILP and linear programming, which belong to optimization techniques where the constraints or objective functions are nonlinear functions. But it may not guarantee the solution to optimization. In the case of linear programming, particularly problems of nonlinear programming, HEMs is not common.

***Mixed-integer nonlinear programming*:** Mixed-integer nonlinear programming (MINLP) is also an optimization technique in which the constraints or objective function provided is a nonlinear function that contains mixed variables. MINLP has been effectively applied to HEMs.

***Dynamic programming*:** In the case of dynamic programming (DP), the total problem is categorized into different subproblems, such as the original problem, which is solved via simpler problems. The optimal solutions of the subproblems are usually kept in time, which allows prior calculations to assist in the next steps. DP can be applied effectively to HEMs.

***Stochastic programming*:** Stochastic programming is also an appropriate optimization technique where the constraints and objective function are uncertainties in variables and parameters. The optimal value of the objective is fed by the predictable value of the objective function. Stochastic programming is used where knowledge regarding the probability distribution function is available. Because of its properties, stochastic programming can be applied effectively to HEMs, not only as a single process but also linked with other processes to estimate and deal with uncertainty.

Robust programming: Like in stochastic programming, in robust programming, uncertainty is treated as variables and parameters. It has no knowledge or idea regarding the function of the probability distribution parameters that are associated with the uncertainty. Despite many demerits, robust programming has been successfully applied to HEM_S.

Model predictive control: Model predictive control (MPC) is a modern, advanced process of control based on the receding horizon principle. Its major aim is to estimate the best course of action at the time of meeting the necessities. Today, MPC is applied significantly to HEMs.

5.9 CONCLUSION

This study must be effective to improve the performance and design of novel HEMs with superior properties without affecting the sustainable environment because of the development of materials synthesized via green technology that minimizes environmental contamination. Today, emphasis is given to developing environmentally friendly HEMs via the pathway of green technology without lead and chlorine that can be used to produce explosives and propellants. The use of energetic polymers, biopolymers, and their composites enhances the performance of the HEMs with adequate processability and mechanical properties. Polymeric materials having excellent performance and mechanical properties that require no oxidizer and undergo detonation or deflagration very quickly as compared to metal-based materials are termed as energetic polymers. This study provides modifications to the performance and formation of novel HEMs like HEDMs, fuel materials, energetic polymer composites, plasticizers, and oxidizers. The modified HEMs are highly essential for spacecraft and the defence sector. The low-susceptibility ammunitions with potentially high performance and spin-offs are the advantages of the proposed study of the rocket propellants, which are now formulated and designed on a priority basis. The most used HEMs include DHT, BTATz, Azineare, DAAT, etc., which are highly stable, have high enthalpy of formation, generate huge volumes of N_2 owing to combustion, and are relatively more environmentally friendly than traditional explosive materials. To improve and modify the defense sector, HEMs are highly effective. Although many works are now reported relating to HEMs, more research is essential for the development and design of superior-performance, eco-friendly explosives and propellants that can be used both in defense and civil works. The high-energy HTPB propellant shows outstanding combustion performance, nontoxic combustion products, low infrared radiation, adequate mechanical performance, but the major disadvantage is high risk of detonation [79–81].

KEYWORDS

- **high-energy materials**
- **biodegradable**
- **eco-friendly**
- **pyrotechnics**
- **sustainable**
- **high energy density materials**

REFERENCES

1. Talawar, M.B., Sivabalan, R., Anniyappan, M., Gore, G.M., Asthana, S.N., Gandhe, B.R. Emerging trends in advanced high energy materials. *Combustion, Explosion, and Shock Waves* 2007, 43(1), 62–72.
2. Martin, W.R., Ball, D.W. Small organic fulminates as high energy materials. Fulminates of acetylene, ethylene, and allene. *Journal of Energetic Materials* 2019, 37(1), 70–39.
3. Aguero, S., Terreux, R. Degradation of high energy materials using biological reduction: A rational way to reach bioremediation. *International Journal of Molecular Sciences* 2019, 20(22), 5556–5578.
4. Gundawar, M.K. Chapter 18—high-energy materials application, in: *Laser-Induced Breakdown Spectroscopy*, 2nd edn., Elsevier Science 2020, pp. 401–419.
5. Chaturvedi, S., Dave, N.P. Solid propellants: AP/HTPB composite propellants. *Arabian Journal of Chemistry* 2019, 12(8), 2061–2068.
6. DeLuca, L. Innovative solid formulations for rocket propulsion. *Eurasian Chemico-. Technological Journal* 2016, 18(3), 181–196.
7. Ugrčić, M. Numerical simulation of the fragmentation process of high explosive projectiles. *Scientific Technical Review* 2013, 63(2), 47–57.
8. Nair, U.R., Asthana, S.N., Rao, A.S., Gandhe, B.R. Advances in high energy materials. *Defence Science Journal* 2010, 60(2), 137–151.
9. Adam, S. Cumming New trends in advanced high energy materials. *Journal of Aerospace Technology and Management* 2009, 1(2), 161–166.
10. Bobrovnikov, S.M., Gorlov, E.V., Zharkov, V.I. Remote detection of traces of high-energy materials on an ideal substrate using the Raman effect. *Atmospheric Oceanic Optics* 2017, 30, 604–608.

11. Butler, J.K. Review of chemistry of high-energy materials, 2nd ed. *Journal of Chemical Education* 2014, 91(2), 163–164.
12. Saikia, A.S., Gore, R., M.ikder, G.A. Synthesis of some potential high energy materials using metal nitrates; An approach towards environmental benign process. *Journal of Scientific and Industrial Research* 2014, 73(7), 485–488.
13. Wang, Y., Liu, Y., Song, S. et al. Accelerating the discovery of insensitive high-energy-density materials by a materials genome approach. *Nature Communication* 2018, 9, 2444–2455.
14. Zhang, W., Zhang, J., Deng, M., Qi, X., Nie, F., Zhang, Q. A promising high-energy-density material. *Nature Communication* 2017, 8(1), 181–188.
15. Agrawal, J.P. Some new high energy materials and their formulations for specialized applications. *Propellants, Explosives, Pyrotechnics* 2005, 30(5), 316–328
16. Yongjin, C., Shuhong, B. High Energy Density Material (HEDM)—progress in research azine energetic compounds. *Johnson Matthey Technology Review* 2019, 63(1), 51–72.
17. Németh, K., Kósik, S. Review of explosive hydrovolcanism. *Geosciences,* 2020, 10(2), 44–72.
18. Li, X., Cao, W., Song, Q. Study on energy output characteristics of explosives containing B/Al in the air blast. *Combustion Explosive Shock Waves* 2019, 55, 723–731.
19. Rocco, D.D., Pasquier, M., Albrecht, E., Carron, P.-N., Dam, F. HEMS inter-facility transfer: A case-mix analysis. *BMC Emergency Medicine* 2018, 18, 13–20.
20. Juniot, W.L.R.,Borges, F.A.S., Rabelo, R.D.A.L., Rodrigues, J.J.P.C., Fernandes, R.A.S., da Silva, I.N. *International Journal of Energy Research*. 2020, Special Issue Research Article, 1–17. https://doi.org/10.1002/er.5183.
21. Beyazkilic, P., Yildirim, A., Bayindir, M. Formation of pyrene excimers in mesoporous ormosil thin films for visual detection of nitro-explosives. *ACS Applied Materials & Interfaces* 2014, 6(7), 4997–5004.
22. Lease, N., Kay, L.M., Brown, G.W., Chavez, D. E; Leon, P., Robbins, D., Manner, V.W. Modifying nitrate ester sensitivity properties using explosive isomers. *Crystal Growth & Design* 2019, 19(11), 6708–6714.
23. Gao, Y., Xu, W., Zhu, D., Chen, L., Fu, Y., He, Q., Caoa, H., Cheng, J. Highly efficient nitrate ester explosive vapor probe based on multiple triphenylaminopyrenyl-substituted. *POSS, Journal of Materials Chemistry A* 2015, 3, 4820–4826,
24. Yang, J., Wang, G., Gong, X., Zhang, J., Wang Yan, A. High-energy nitramine explosives: A design strategy from linear to cyclic to caged molecules. *ACS Omega* 2018, 3, 9739–9745
25. Elbeih, A., Zeman, S., Jungová, M., Vávra, P. Attractive nitramines and related PBXs. *Propellants, Explosives, Pyrotechnics* 2013, 38(3), 379–385.
26. Shreeve, J.M., Zhang, J., Yin, P., Mitchell, L.A., Parrish, D.A. N-Functionalized nitroxy/azido fused-ring azoles as high-performance energetic materials. *Journal of Materials Chemistry A* 2016, 4(19), 7430–7436.
27. Sabatini, J.J., Oyler, K.D. Recent advances in the synthesis of high explosive materials. *Crystals* 2016, 6(1), 5–27.

28. Chaudhary, S., Sonkusre, P., Bhasin, K.K., Sabherwalb, P., Suri, R. Trace detection of some nitro-explosives using thermal mediated immunochemical defragmented method. *Biosensors and Bioelectronics* 2019, 126(1), 590–595.
29. Horie, Y. Hot spots, high explosives ignition, and material microstructure. *Materials Science Forum* 2013, 767, 3–12.
30. Capka, L., Vecera, Z., Mikuska, P., Sesták, J., Kahlea, V., Bumbová, A. A portable device for fast analysis of explosives in the environment. *Journal of Chromatography A* 2015, 138, 167–173.
31. Wu, B.-D., Zhou, Z.-N., Li, F.-G., Yang, L., Zhang, T.-L., Zhang, J.-G. Preparation, crystal structures, thermal decompositions and explosive properties of two new high-nitrogen azideethylenediamine energetic compounds. *New Journal of Chemistry* 2013, 37, 646–653.
32. Behera, R., Biswal, T., Panda, R.B. Recent progress in explosives: A brief review, in: S.K. Acharya, D.P. Mishra (Eds.), *Current Advances in Mechanical Engineering. Lecture Notes in Mechanical Engineering*. Springer: Singapore, 2021, https://doi.org/10.1007/978-981-33-4795-3_29.
33. Zhang, K., Ni, O. Rheological properties and stability of emulsion explosive matrix. *Journal of Dispersion Science and Technology* 2015, 37(7), 932–937.
34. Oyler, K.D. Green primary explosives. In: *Green Energetic Materials*. John Wiley & Sons, Ltd., 2014, pp. 103–132.
35. Huynh, M.H.V., Coburn, M.D., Meyer, T.J., Wetzler, M. Green primary explosives: 5-Nitrotetrazolato-N2-ferrate hierarchies. *Proceedings of the National Academy of Sciences* 2006, 103(27), 10322–10327.
36. Li, Y., Huang, H., Shi, Y., Yang, J., Pan, R., Lin, X. Potassium nitraminofurazan derivatives: Potential green primary explosives with high energy and comparable low friction sensitivities. *Chemistry—A European Journal* 2017, 27(30), 7353–7360.
37. Zhang, Q., Chen, D., Jing, D., Fan, G., Li, L., He, H., Wang, W., Niea, F. Access to green primary explosives via constructing coordination polymers based on bis-tetrazole oxide and non-lead metals. *Green Chemistry* 2019, 21(8), 1947–1955.
38. Ernst-Christian, K. Insensitive high explosives: V. ballistic properties and vulnerability of nitroguanidine based propellants. *Propellants, Explosives, Pyrotechnics* 2021, 46(2), 174–206.
39. Kudryavtsev, P. Production technology development and creation of production of additives used in solid rocket propellants. *Journal "Scientific Israel—Technological Advantages"* 2014, 16(3), 25–37.
40. Bird, D.T., Ravindra, N.M. A review: Advances and modernization in U.S army gun propellants. *JOM* 2021, 73, 1144–1164.
41. Druet, L., Asselin, M. A review of stability test methods for gun and mortar propellants, i: The chemistry of propellant ageing. *Journal of Energetic Materials* 1988, 6(1–2), 27–43.
42. Kubo, S., Fuku, Y., Shimamoto, T., Kuwayama, A. Ohya, M., Amano, H. Effect of ageing on the combustion of single base propellants. *Combustion and Flame* 2020, 221, 212–218.
43. Xiufang, M., Weihua, Z., Jijun, X., Heming, X. Molecular dynamics study of the structure and performance of simple and double bases propellants 2008, 156(1–3), 201–207.

44. Ghosh, K., Pant, Chandra, S., Sanghavi, R., Adhav, S., Singh, A. Studies on triple base gun propellant based on two energetic azido esters. *Journal of Energetic Materials* 2008, 27(1), 40–50.
45. Landsem, E., Jensen, T.L., Hansen, F.K., Unneberg, E., Kristensen, T.E. Mechanical properties of smokeless composite rocket propellants based on prilled ammonium dinitramide. *Propellants, Explosives, Pyrotechnics* 2012, 37(6), 691–698.
46. Hegab, A.M., Gutub, S.A. An overview of composite propellant burning. *International Journal of Engineering Trends and Technology* 2015, 27(2), 118–123.
47. Davenas, A. Development of modern solid propellants. *Journal of Propulsion and Power* 2003, 19(6), 1108–1128.
48. Mason, B.P., Roland, C.M. Solid propellants. *Rubber Chemistry and Technology* 2019, 92(1), 1–24.
49. Mehendiratta, D., Ramachandran, M. A review on different propellant materials for space vehicles and their characterization. *International Journal of Mechanical and Production Engineering Research and Development* 2018, Special Issue, 93–102.
50. Damse, R.S., Singh, A., Singh, H. High energy propellants for advanced gun ammunition based on RDX, GAP and TAGN compositions. *Propellants, Explosives, Pyrotechnics* 2007, 32(1), 52–60.
51. Okninski, A. Solid rocket propulsion technology for de-orbiting spacecraft. *Chinese Journal of Aeronautics* 2022, 35(3), 128–154.
52. Trache, D., Klapötke, T.M., Maiz, L., Abd-Elghany, M., DeLuca, L.T. Recent advances in new oxidizers for solid rocket propulsion. *Green Chemistry* 2017, 19(20), 4711–4736.
53. Chandru, R.A., Balasubramanian, N., Oommen, C., Raghunandan, B.N. Additive manufacturing of solid rocket propellant grains. *Journal of Propulsion and Power* 2018, 34(4), 1090–1093.
54. Edwards, T. Liquid fuels and propellants for aerospace propulsion: 1903–2003. *Journal of Propulsion and Powe*r 2003, 19(6), 1089–1107.
55. Zhang, Q., Shreeve, J.M. Ionic liquid propellants: Future fuels for space propulsion. *Chemistry—A European Journal* 2013, 19(46), 15446–15451.
56. Thirupathi, M., Madhavi, N., Naidu, K.S. Design and analysis of a fuel injector of a liquid rocket engine. *International Journal of Engineering and Advanced Technology* 2015, 4(5), 223–229
57. Crocco, L. Aspects of combustion stability in liquid propellant rocket motors part I: Fundamentals. Low frequency instability with monopropellants. *Journal of the American Rocket Society* 2012, 21(6), 163–178.
58. Lee, K.H. Numerical comparison of exhaust plume flow behaviors of small monopropellant and bipropellant thrusters. *PLoS One* 2017, 12(5), 1–20.
59. Rong, L., Man, Y., Lv, Y., Zou, S. *[IEEE 2020 39th Chinese Control Conference (CCC)—Shenyang, China (2020.7.27-2020.7.29)] 2020 39th Chinese Control Conference (CCC)—Tripropellant Hypersonic Wind Tunnel Modeling and Switched Finite-time Control*, 2020, pp. 2101–2106. doi:10.23919/CCC50068.2020.9188898.
60. Pérez-Roca, S., Marzat, J., Piet-Lahanier, H., Langlois, N., Farago, F., Galeotta, M., Le Gonidec, S. A survey of automatic control methods for liquid-propellant rocket engines. *Progress in Aerospace Sciences* 2019, 107, 63–84.

61. Padwa, M.B., Benveniste, N., Mishra, D.P. Gel propellants. *Progress in Energy and Combustion Science* 2021, 80, 1–150.
62. Siva, V. Theoretical investigation of liquid propellant rocket engine performance. *International Journal of Engineering Research and Application* 2018, 8(3), 38–41.
63. Kang, Z., Wang, Z., Li, Q., Cheng, P. Review on pressure swirl injector in liquid rocket engine. *Acta Astronautica* 2018, 145, 174–198.
64. Janardan, B.A., Daniel, B.R., Zinn, B.T. Measurements of acoustic responses of gaseous propellant injectors. *Journal of Sound and Vibration* 1976, 47(4), 559–569.
65. Arkhipov, V.A., Zarko, V.E., Zharova, I.K. Solid propellant combustion in a high-velocity crossflow of gases (review). *Combustion, Explosion, and Shock Waves* 2016, 52, 497–513
66. Conkling, J.A., Mocella, C.J. *Chemistry of Pyrotechnics: Basic Principles and Theory*, 3rd edn., CRC Press, 2018, p. 360, https://doi.org/10.1201/9780429262135.
67. Steinhauser, G., Klapötke, T.M. "Green" pyrotechnics: A chemists' challenge. *Pyrotechnics* 2008, 47(18), 3330–3347.
68. Danali, S.M., Palaiah, R.S., Raha, K.C. Developments in pyrotechnics. *Defence Science Journal* 2010, 60(2), 152–158.
69. Zou, D. Explosives, in: *Theory and Technology of Rock Excavation for Civil Engineering*. Springer: Singapore. Chapter 3, 2017, pp. 105–170. https://doi.org/10.1007/978-981-10-1989-0.
70. Mathieu, J., Stucki, H. Military high explosives. *CHIMIA International Journal for Chemistry* 2004, 58(58), 383–389.
71. Araos, M., Onederra, I. Development of a novel mining explosive formulation to eliminate nitrogen oxide fumes. *Mining Technology* 2015, 124(1), 16–23.
72. Afiq, M., Huri, M., Ahmad, U., Ibrahim, R., Omar, M. A review of explosive residue detection from forensic chemistry perspective. *Malaysian Journal of Analytical Sciences* 2017, 21(2), 267 – 282.
73. Sanchidrián, J.A., López, L.M., Segarra, P. The influence of some blasting techniques on the probability of ignition of firedamp by permissible explosives 2008, 155(3), 580–589.
74. Gruau, C., Picart, D., Belmas, R., Bouton, E., Delmaire-Sizes, F. Ignition of a confined high explosive under low velocity impact. *International Journal of Impact Engineering* 2009, 36(4), 537–550.
75. Zygmunt, A., Gańczyk., K., Kasztankiewicz., A., Cieślak., K., Goło, T. Application and properties of aluminum in primary and secondary explosives. *Journal of Elementology* 2017, 22(2), 747–759.
76. Bondarchuk, S.V. Modeling of explosives: 1,4,2,3,5,6-dioxatetrazinane as a new green energetic material with enhanced performance. *Journal of Physics and Chemistry of Solids* 2020, 142, 1–8.
77. Lima, D.R.S., Bezerra, M.L.S., Neves, E.B., Moreira, F.R. Impact of ammunition and military explosives on human health and the environment. *Reviews on Environmental Health* 2011, 26(2), 101–110.
78. Zhou, L.Y., Zu, W. Dou, X. Qualitative detection toward military and improvised explosive vapors by a facile TiO_2 nanosheet-based chemiresistive sensor array. *Frontiers in Chemistry* 2020, 8, 29–41.

79. Gomes, I., Bot, K., Ruano, M.G., Ruano, A. Recent techniques used in home energy management systems: A review. *Energies* 2022, *15*(8), 2866–2906.
80. Amiri, A., Shahbazian-Yassar, R. Recent progress of high-entropy materials for energy storage and conversion. *Journal of Materials Chemistry A* 2021, *9*(2), 782–823.
81. Lei, Z., Liu, X., Wang, H., Wu, Y., Jiang, S., Lu, Z. Development of advanced materials via entropy engineering. *Scripta Materialia* 2019, *165*, 164–169.

CHAPTER 6

Machine Learning-Based Green and Energy Efficient Architecture for Next Generation Cellular Networks

DEEPA NAIK* and TANMAY DE

Department of Computer Science and Engineering, National Institute of Technology, Durgapur, West Bengal, India

*Corresponding author

ABSTRACT

The demand for broadband spectrum has increased due to the needs of these areas and metropolitan civilizations. Despite the enormous bandwidth potential of optical networks, it is difficult and expensive to penetrate these areas due to the harsh terrain. WiMAX (Worldwide Interoperability for Microwave Access) is therefore used to provide consumers with a front-end solution. Furthermore, the infrastructure is poor and the power supply is unstable in these isolated, rocky terrains. Alternative energy sources can help us in solving the issue of unstable electricity. To forecast renewable energy in the region where the network node is located, machine learning is applied. Nonrenewable energy sources are used to reduce the unpredictable nature of renewable energy. Eighty percent of the total power used in wired networks is consumed by the access segment. Utilizing renewable energy sources can reduce the consumption of brown energy. Renewable energy comes from sources including geothermal heat, water, wind, sunlight, and tides. Energy is heavily used by optical network units in passive optical networks and base stations in wireless access networks. In integrated optical-wireless access networks, traffic grooming/aggregations

Advances in Energy Materials: Materials Fabrication, Properties and Applications.
Lishin T, Maciej Jaroszewski, and Sabu Thomas (Eds.)

DOI: 10.1201/9781779643674-6

reduce the use of nonrenewable energy. The major goal is to use as much renewable energy as possible while using as minimum amount of brown energy as possible to meet traffic demands.

6.1 INTRODUCTION

The integration of the optical network with the wireless network helps the network operator to deploy the network in a cost-efficient way. The optical fibers have high capacity, reliability, and latency [1]. The integration of the optical network with several wireless technologies such as WiFi, WiMAX, or long-term evaluation increases coverage area. Depending on the coverage area needed to be deployed, these wireless technologies are integrated with the passive optical networks [2].

Most previous studies on hybrid access networks focused on architecture design, survivability, and guaranteed quality of service (QoS), as well as their role in other critical infrastructures, like smart grids [3]. However, the energy efficiency aspect of hybrid access networks, only considers the energy-saving aspects in the passive network, while giving less attention to wireless networks. The energy-saving in wireless devices also reduces the operational cost of the network operators. Meanwhile, access networks consume more energy than required to meet actual traffic loads. The energy efficiency solutions for optical backhaul networks have been intensively explored. Backhaul networks account for up to 50% of overall power usage in communication networks [4]. Despite this, access network devices are underutilized and consequently stay inactive for the majority of the time [5], resulting in a significant amount of lost energy.

Power-saving modes of routers are discussed in [6–11] and they are based on optical network unit (ONU) sleep mode [6, 7]. In these cases, traffic demands are routed through alternate routes by making some ONUs sleep. Both power saving for wireless devices and ONUs have been studied in [8–11]. However, these and other existing solutions do not look into the details of how to incorporate the power-saving modes into the dynamic bandwidth allocation process in the optical backhaul, decreasing the efficiency of network resource management.

In this chapter, we propose a novel energy conservation scheme for a WDM passive optical networks (PON)-based wireless network using the renewable resources. It aims at jointly scheduling power-saving modes for not only wireless devices (i.e., relay stations and access points (APs)) in the wireless front-end but also ONUs in the optical backhaul in a coordinated and synchronized manner to minimize the total energy consumption. Instead of minimizing the number of active ONUs, our algorithm exploits the idle time of ONUs and puts them in sleep mode along with their integrated AP to save energy. In the WDM PON-based hybrid network, each ONU is assigned the dedicated wavelength. The ONU-AP schedules transmission subplots for its associated relay station in a TDMA manner, thus allowing them to enter seep mode outside their subplots to minimize the energy consumption of the AP. Transmission scheduling using TDMA eliminates collisions [12].

The main contributions of this chapter are as follows:

1. The traffic grooming approach is applied for a WDM PON-based wireless network, to utilize the network resources efficiently.
2. Minimize the overall energy consumption without compromising the delay performance of integrated hybrid access networks.
3. To save energy for all ONUs, APs, and relay nodes by using a synchronization and incorporation scheme to schedule bandwidth for each relay station.
4. The renewable energy resources used to serve the traffic demands.

Here the hybrid network architecture and the energy-saving scheme are explained in Section 6.2. The energy-saving model for the hybrid network is presented in Section 6.3. Section 6.4 presents the reports on results and obtained findings. Finally, conclusions are drawn in Section 6.5.

6.2 ENERGY SAVING SCHEME FOR HYBRID NETWORKS

6.2.1 HYBRID NETWORK ARCHITECTURE

Figure 6.1 depicts the hybrid network architecture, which is based on WDM PON and WLAN technologies. The optical network consists of a point-to-multipoint tree type of topology. The connectivity between ONUs and arrayed waveguide gratings is done by optical line terminal (OLTs). Each ONU was assigned a dedicated wavelength depending on the traffic demands. The OLT and an ONUs are separated by a distance of 10–20 km. The ONUs are housed at subscriber locations (both business and personal). The ONUs may be used by a single customer or by different customers.

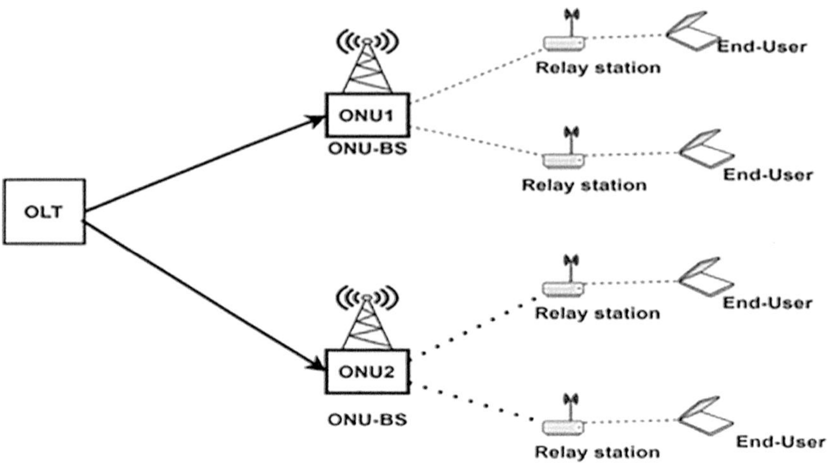

FIGURE 6.1 Hybrid network architecture.

6.2.2 *PROBLEM STATEMENT*

An energy-saving model is developed to provide operators with an overview of parameters and quantities that determine the potential energy saving for a given network configuration. The traffic initially traverses the renewable energy-aware routes to route the traffic. Over a period of time, due to the network load the traffic switches to the nonrenewable energy routes to provide uninterrupted service to the customers.

6.3 PROPOSED ALGORITHM FOR HYBRID NETWORK

This chapter proposes a novel energy-saving scheme for WDM PON-based wireless networks. The main objective is to jointly optimize the energy-saving mechanism for both the optical and wireless networks. The wireless devices are turned off and on depending on the operation of the ONU-AP. When allocating the bandwidth, the OLT to ONU-AP uses the medium access control (MAC) layer because of the point-to-point connection between the OLT and ONUs. This MAC protocol leads to collision-free communication between the OLT and other ONUs.

The OLT transmitter can enter into sleep mode once the wavelength is scheduled to ONUs. The ONU-AP) to the wireless network uses the

TDMA time slots method. ONU-AP employs TDMA scheduling to assign an upstream subslot for each relay station in the next cycle based on its requested bandwidth and broadcasts downstream frames to all relay stations. In this work, the relay station transmitter and receiver are assumed to enter sleep mode independently. The idle ONU-APs which are outside their time slot put them to sleep mode, along with the associated relay stations. Energy consumed by the hybrid network is the summation of the energy consumed by the optical network and the energy consumed by the wireless network.

PROPOSED ALGORITHM

Step 1: The machine learning technology is utilized to detect the nodes that work on green energy. Every network node has nonrenewable energy available as a backup.

Step 2: A pair of matrices is created by splitting the traffic matrix. The first matrix indicates the quantity of requests that use up the wavelength channel's entire capacity. For this, transponders at sender and receiver are required, together with the full wavelength channel.

Step 3: The traffic demands that use up the wavelength channels' partial capacity make up the second matrix. These requests can be tailored in conjunction with other requests to optimize network use and reduce energy usage. The traffic requests are routed through the transponders that are currently in use.

Step 4: Determine the routes that can be powered by renewable energy between the s-d pairs by selecting the traffic demands from the first matrix. In the event that the nodes are not powered by renewable energy, meet traffic demands with nonrenewable energy.

Step 5: The traffic demands not served are put into a blocked list. Put the network components in sleep mode once their turn completes. And wake up the network components before their turn restart. Finally, calculate the energy consumed in the given network, when sleep mode is activated in the network.

Network nodes out of their time slot are put in sleep mode, leading to huge savings in the energy consumption compared to normal mode (where all the network components are kept active). The traffic aggregation techniques along with the sleep mode operations have great contributions in energy savings.

The notations used in this literature are given in Table 6.1.

TABLE 6.1 Notations and Descriptions

Notation	Description	Value
$\text{Time}_{\text{Cycle}}$	Passive optical network polling time	Seconds
$\text{Time}_{\text{Solts}}$	Slots allocated to each ONU_AP	Numbers (4, 8)
$\text{Time}_{\text{Guard}}$	Guard time between the consecutive slots	1 μs
$\text{Time}_{(\text{Sleep-to-Wakeup})}$	Time takes to activate device ONU, AP, Relay station	2 ms
$P_{\text{ONU\_}}{}^{\text{Active}}/P_{\text{ONU}}{}^{\text{sleep}}$	Power consumed by ONU during active/sleep (also for AP)	500 W, 50 W
$P_{\text{Relay}}{}^{\text{active}}/P_{\text{Relay}}{}^{\text{sleep}}$	Power consumed by relay station during active/sleep	300 W, 300 W
M, N	Number of ONU and Relay station	8,16
Wavelengths (λ)	40 Gb/s	8
$P_{\text{Router port (40 Gb/s)}}$	40 Gb/s	600 W

The energy consumption in the traditional scheme is denoted below.

In the traditional method, the ONUs are turned off during the idle period. The wireless devices are kept active.

$$E = \text{Time}_{\text{cycle}}\ (P_{\text{ONU}}{}^{\text{active}} + P_{\text{AP}}{}^{\text{active}}) + N * P_{\text{Relay}}{}^{\text{active}} \tag{6.1}$$

$$\text{where Time}_{\text{cycle}} = M * (\text{Time}_{\text{slots}} + \text{Time}_{\text{Guard}}) \tag{6.2}$$

where timeslot is the average time slots allocated to each ONU, $\text{Time}_{\text{Guard}}$ is the guard band, and M is the number of ONU in the network, N is the number of relay stations in the wireless network.

The energy consumption in the proposed model is given below.

In the proposed method, we have considered the energy saving in the optical and the wireless domains by turning off the network devices that are outside their time slots. This is achieved by time synchronization between the OLT and ONU-AP in the optical domain and ONU-AP to relay station time synchronization in the wireless domain. This leads to overall energy minimization in the hybrid network.

$$E\_optical = P_{ONU}{}^{active}*(Time_{slot} + Time_{(sleep\_wakeup)}) + P_{AP}{}^{active}(Time_{slot} + Time_{(sleep\_wakeup)}) + P_{ONU}{}^{sleep}(Time_{cycle} - Time_{slot} - Time_{(Sleep\text{-}to\text{-}Wakeup)}) + P_{AP\_}{}^{sleep}(Time_{cycle} - Time_{slot} - Time_{(Sleep\text{-}to\text{-}Wakeup)}) \quad (6.3)$$

$$E1\_Wireless = N*P_{Relay}{}^{Receivers} + P_{Realy}{}^{transmitter} + N*P_{Realy}{}^{transmitter\_(Time\ sleep\text{-}to\text{-}wakeup)} \quad (6.4)$$

The energy consumed during the "*N*" active transmitters and receivers is given by the above formula. Only one transmitter is active, so the power consumption of the active transmitter is considered.

Outside the timeslot, all relay station receivers and transmitters are put into sleep mode and each of the relay receivers wakes up once immediately before the time slot for receiving downstream broadcast frames.

$$E2\_Wireless = N*P_{relay}{}^{sleep}(Time_{cycle\text{-}}Time(_{slot\text{-}to\text{-}wakeup}) + N*(P_{relay\_}{}^{receiver(sleep\text{-}to\text{-}wakeup)}) \quad (6.5)$$

The efficiency achieved in the proposed algorithm is given by

$$Efficiency = (E - E\_Optical - E1\_Wireless - E2\_Wireless)/E \quad (6.6)$$

6.4 RESULT ANALYSIS

In this chapter, we have considered solar and wind energy as the source of renewable energy for a given network. The nonrenewable energy is provided as the backup energy. The machine learning technology is utilized to detect the nodes that are powered by renewable energy sources. The nonrenewable energy consumption at the node is zero, when the node in the network has enough renewable energy to serve the traffic.

The renewable source of solar energy calculation formula is

$$(1 - skycover)At \quad (6.7)$$

We assume that one square meter silicon solar cell can produce a daily average of 0.28 kW. A is the spread of solar panels and t is the time. $0.28 \times (1 - 50\%) \times 1 \times 1 = 0.14$ kWh power on that day.

The renewable source of wind energy calculation formula is

$$1/2 \times \rho \times A \times v^3 \times t \quad (6.8)$$

where ρ is the air density (in kg/m³), A is the airflow area, V is the velocity of the wind speed, and t is time in seconds.

The power in the wind at 5 m/s is: $1/2 \times \rho \times A \times v^3 = 0.5 \times 1.225$ kg/m³ $\times$ 452.4 m² $\times$ (5 m/s)³ $\times$ 1 = 34,636 W = 34.63 kWh.

Table 6.2 depicts the harvested renewable energy at each network node from May 1 to May 4. Accurate decisions are made and complicated relationships or patterns are developed from empirical data with the aid of machine learning algorithms [13]. Multilayer perceptron models and support vector machines (SVM) are shown to be useful for studying a system's future characteristics. As a result, we have successfully predicted weather forecast reports using ANN and SVM algorithms. The following settings are used to train the multilayer perceptron model: four hidden layers and four hidden neurons. In training data, this yielded an accuracy of 0.93%, while SVM yielded 0.86% accuracy.

TABLE 6.2 Renewable Energy Outputs from May 1, 2018 to May 8

Day	Sky Cover	Solar Energy (kWh)	Wind Speed (m/s)	Wind Energy (kWh)	Renewable Energy (Solar + Wind) (kWh)
May 1	40	483	5	276	759
May 2	50	403	6	478	881
May 3	20	645	7	760	1405
May 4	30	564	4	141	705

Table 6.3 depicts the comparison between the nonrenewable energy consumption with nonrenewable energy consumption with sleep mechanism. The sleep mechanism leads to more effective in saving the energy.

TABLE 6.3 Comparison of Energy Consumption in the Given Network

Day	Nonrenewable Energy Consumption (kWh)	Nonrenewable Energy Consumption with Sleep Mechanism (kWh)
May 1	400	286
May 2	462	297
May 3	483	456
May 4	389	340
May 5	1200	1011

From Figure 6.2 observation it is concluded that the energy saving depends on many variables, such as renewable energy harvested at the node, including cycle time, which is a function of the network traffic load, number of ONUs, number of relay stations per ONU-AP, power profile of the network devices, and their waking-up overhead times. The devices that are outside the time slots are switched off leading to energy saving without compromising the QoS related to the traffic demands in the upstream as well as downstream direction.

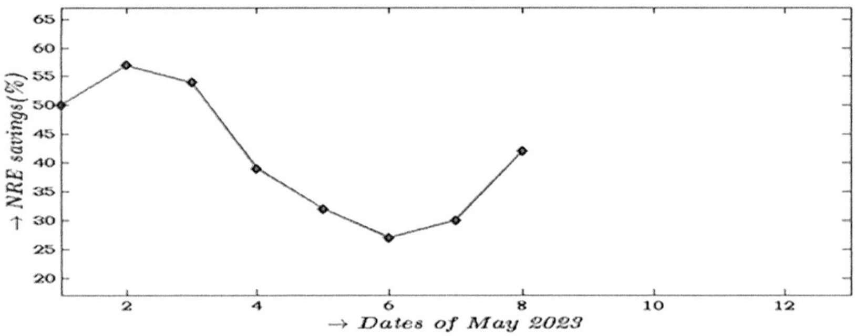

FIGURE 6.2 Variation in renewable energy leads to variation in the nonrenewable energy savings.

Figure 6.3 depicts the comparisons between nonrenewable energy savings with sleep mode and without sleep mode for hybrid networks. During the period of May 3 and May 4 the energy savings are low due the heavy traffic at the network nodes. Both algorithms performed the same results.

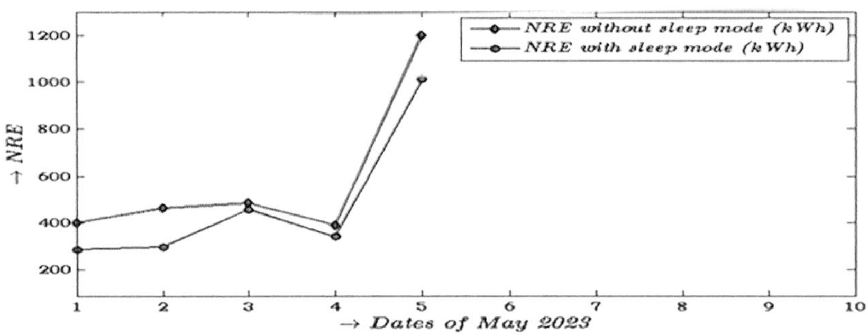

FIGURE 6.3 Nonrenewable energy savings with sleep mode and without sleep mode for hybrid networks.

6.5 CONCLUSION

This chapter proposed an energy-saving scheme for WDP-PON-based wireless access networks. The existing algorithms consider the energy-saving scheme for optical and wireless networks in an isolated way. However, this is not the optimal solution for handling the hybrid type of network. Unlike existing solutions, bandwidth allocation and power-saving mode scheduling of ONUs, AP, and relay stations are jointly performed in the hybrid networks to improve overall energy efficiency and network performance. The obtained results confirm that it is beneficial to apply TDMA to wireless networks for improving energy efficiency and decreasing delay. By exploiting the idleness of network components outside their transmission slots for power-saving modes, the proposed scheme achieves significant energy-saving while maintaining other QoS.

KEYWORDS

- **base station**
- **long-term evolution**
- **optical line terminal**
- **optical network unit**
- **wireless fidelity**

REFERENCES

1. Beyranvand, H., Lim, W., Maier, M., Verikoukis, C., Salehi, J. Backhaul-aware user association in FiWi enhanced LTE—A heterogeneous networks. *IEEE Transactions Wireless Communication* 2015, 14(6), 2992–3003.
2. Aurzada, F., L´evesque, M., Maier, M., Reisslein, M. FiWi access networks based on next-generation PON and gigabit-class WLAN technologies: A capacity and delay analysis. *IEEE/ACM Transactions on Networking* 2014, 22(4), 1176–1189.
3. Maier, M., L´evesque, M. Dependable fiber-wireless (FiWi) access networks and their role in a sustainable third industrial revolution economy. *IEEE Transactions* 2014, 63(2), 386–400.

4. Tombaz, S., Monti, P., Farias, F., Fiorani, M., Wosinska, L., Zander, J. Is backhaul becoming a bottleneck for green wireless access networks? in: *Proceedings of IEEE ICC*, 2014, pp. 4029–4035.
5. Valcarenghi, L., Pham Van, D., Raponi, P.G., Castoldi, P., Campelo, D., Wong, S., Yen, S., Kazovsky, L.G., Yamashita, S. Energy efficient in passive optical networks: Where, when, and how? *IEEE Network* 2012, 26(6), 61–68.
6. Chowdhury, P., Tornatore, M., Sarkar, S., Mukherjee, B. Building agreen wireless-optical broadband access network (WOBAN). *Journal of Lightwave Technology* 2010, 28(16), 2219–2229.
7. Kantarci, B., Mouftah, H.T. Energy efficiency in the extended-reach fiber-wireless access networks. *IEEE Network* 2012, 26(2), 28–35.
8. Miyanabe, K., Nishiyama, H., Kato, N., Ujikawa, H., Suzuki, K.-I., Yoshimoto, N. Synchronized power saving mechanisms for battery powered mobile terminals in smart FiWi networks. in: *Proceedings of IEEE VTC*, 2014, pp. 1–5.
9. Barradas, A., Correia, N., Coimbra, J., Sch¨utz, G. Load adaptive and fault tolerant framework for energy saving in fiber-wireless access networks. *Journal of Optical Communications and Networking* 2013, 5(9), 957–967.
10. Coimbra, I.J., Schutz, G., Correia, N. Energy efficient routingalgorithm for fiber-wireless access networks: A network formation game approach. *Computer Network* 2014 60, 201–216.
11. Ahmed, M., Ahmad, I., Habibi, D. Service class resource management for green wireless-optical broadband access networks (WOBAN). *IEEE/OSA Journal of Lightwave Technology* 2015 33(1), 7–18.
12. Tsao, S.-L., Huang, C.-H. A survey of energy efficient MAC protocols for IEEE 802.11 WLAN. *Computer Communication* 2011, 34(1), 54–67.
13. Maimó, L.F., Gómez, Á.L.P, Clemente, F.J.G., Pérez, M.G., Pérez, G.M. A self-adaptive deep learning-based system for anomaly detection in 5G networks. *IEEE Access* 2018, *6*, 7700–7712.

CHAPTER 7

Energy and Exergy Analysis of a Solar Thermal Collector Using Two Scheffler Dish Concentrators

SUBHRA DAS[1*] and BIBEK BANDOPADHYAY[2]

[1]*Ansyst Consulting, Gurugram, Haryana, India*

[2]*International Institute of Energy Conservation, New Delhi, India*

*Corresponding author

ABSTRACT

Thermal analyses of solar collectors are performed using the first law of thermodynamics which provides information about the quantity of energy output from the collector for a given energy input. The first law analysis does not give any idea about the quality of energy or the available work output from the system. The work done by Carnot in 1824 provided a basis for computing the work output from a heat engine as a function of temperature difference of hot and cold reservoir. Zoran Rant in 1953 suggested that the technical working capacity of a system be referred to as exergy. In this chapter, energy and exergy analysis of a solar thermal collector consisting of two Scheffler dish concentrators reflecting incident solar radiation on one absorber is presented. It presents a mathematical model to compute thermal efficiency of the collector as a function of incident solar radiation, ambient temperature, wind speed, and absorber temperature. These input parameters can be easily recorded without much intervention with the system. The results obtained from the model for the overall loss coefficient and optical efficiency are compared with

Advances in Energy Materials: Materials Fabrication, Properties and Applications.
Lishin T, Maciej Jaroszewski, and Sabu Thomas (Eds.)

DOI: 10.1201/9781779643674-7

experimental data from the site. A difference of 13.8 W m^{-2} K^{-1} and 0.0187 is observed in the overall loss coefficient and optical efficiency, respectively. An average exergy efficiency of 55% is obtained for the collector over the experimental period.

7.1 INTRODUCTION

Energy is defined in classical thermodynamics as the capacity to do work. Man had utilized energy for his well-being ever since he appeared on Earth. Energy used by primitive men was directly or indirectly derived from sun, that is, men used only renewable sources of energy. In today's industrialized societies, the energy requirement is basically derived from nonrenewable sources of energy like coal, oil, and natural gas which will exhaust after a finite time period and also have many adverse effects on the society. Thus, humanity for its support should adopt a source of energy which is renewable, clean, and safe. The most important source of such energy is the solar energy.

The sun is an abundant permanent source of energy. The solar energy intercepted by planet Earth while traversing the Earth's atmosphere gets diluted by attenuation, local weather phenomena, and air pollution. The amount of radiation received on the earth's surface ranges from 100 to 800 W m^{-2}. Solar energy, an abundant source of energy, is a distributed and diffuse form of energy which requires efficient and meticulously designed collectors to collect the energy and convert it into other useful forms, particularly heat.

The solar thermal collectors [1] are heat exchangers that convert solar energy into thermal energy which is transferred to a heat transfer fluid (HTF). The energy, thus, extracted by the HTF is stored in a storage tank for thermal applications. Thermal efficiency of these collectors is defined as the ratio of useful energy gain by the collector to the total insolation on the collector aperture. The collector efficiency decreases with the increase in temperature difference between the collector and its surrounding due to increased convective and radiative heat losses from the collector surface to ambient.

Broadly, solar thermal collectors can be classified into two types: flat plate collectors and concentrating collectors. Flat plate collectors are majorly used for applications that require hot fluids (air, water, or oil) in the range of 40–120°C. On the other hand, concentrating collectors are used for industrial applications delivering hot fluid at very high temperature to an extent of 2000°C. High temperature can be achieved by concentrating solar radiation at a single point or along a line where an absorber carrying HTF can be placed to collects the energy from the sun. Based on the mechanism, concentrating solar thermal collectors can be broadly classified as imaging (line focus and point focus) and nonimaging collectors. The parabolic trough collectors [2], Fresnel lens collectors [3], Scheffler dish-type collectors, paraboloid dish-type collector [4], and central receiver collectors are imaging collectors whereas the compound parabolic collector is a nonimaging concentrating collector [1]. These types of collectors are majorly utilized for industrial applications for steam generation, in solar furnaces, solar thermal power plants, and many other industrial applications which require heat at a very high temperature [5].

Concentrating collectors consists of a concentrator and an absorber which is placed along the focal line (for line focus collector) or at the focus (for point focus collectors) through which HTF flows [6]. These types of collectors use only the beam radiation to generate the required output. Thus, it tracks the sun so that it is incident normally onto the collector surface. The concentrator collects the incident beam radiation and reflects it to the absorber which absorbs the solar radiation and transfers it to the fluid flowing through it. The temperature of the absorber increases until it reaches the stagnation temperature when the entire energy absorbed by the absorber equals the heat losses from its surface [1]. The heat transfer process is associated with heat losses from the absorber surface to the ambient by convection and radiation and optical losses due to concentrator imperfections, material characteristics, and defocusing of the collector.

Energy analysis is a traditional approach to analyze thermal performance of these collectors. It makes use of the first law of thermodynamics from which energy balance of the system is derived. The analysis provides information about the efficient use of energy in the process and helps in identifying the processes parameters that affect the instantaneous efficiency of the system. The energy approach has some limitations. It does not recognize the different quality of energy, that is, does not give any measure to differentiate between the qualities of heat that is delivered by a system at a low temperature to that at very high temperature. The second law analysis was introduced to overcome these limitations of energy analysis.

Modern approach of thermodynamic optimization makes use of the second law analysis which is based on both first and second law of thermodynamics. It introduces the concept of entropy, entropy generation, and exergy. Entropy generation minimization method is widely used by researchers to analyze and optimize system performance. The second law analysis which takes into account the quality of energy serves mostly for the practical estimation and analysis of processes.

7.1.1 ENERGY, ENTROPY, AND EXERGY

The first law of thermodynamics states that [7]: "Energy is conserved. It can neither be created nor destroyed, only can be transformed from one form to another. The transport of any commodity like heat, electric charge, etc., has an associated energy transport."

The second law of thermodynamics introduces another physical quantity: entropy. Entropy may be viewed as the energy carrier in thermal processes and temperature may be regarded as the thermal potential. While energy is conserved, entropy is not. Entropy is generated in irreversible process and is conserved only for an idealized reversible process. The increase in quantity of entropy in a process depreciates energy qualitatively. It may be defined as [8]:

"Entropy is a substance like quantity (like electric charge) which is contained in bodies and which can flow from body to body. It can be created in irreversible processes (burning, electric currents, friction, heat flow etc.) but cannot be destroyed."

The law of balance of entropy can be written as:

Rate of change of entropy content of a body = Net flow of entropy into and out of the body + Creation of entropy inside the body.

Mathematically [8],

$$\frac{dS}{dt} = I_{s,net} + \Pi_s \tag{7.1}$$

$$\Pi_s \geq 0 \tag{7.2}$$

where S is the entropy content of the body, $I_{s,net}$ is the net flux of entropy and $\prod_s$ is the entropy generated inside a body per unit time. Entropy generation in a system degrades the quality of energy output from the system affecting its ability to do work.

Exergy may be defined as the maximum work output that can be derived from a system undergoing reversible process from an initial state to the state of equilibrium with its environment [9] and is inversely proportional to entropy generation. The exergy fraction of an energy source is its maximum work potential. Hence, it is a measure of its *quality*.

Different approaches to second law analysis have been developed over the last several decades and are widely used by engineers to judge and evaluate the performance of a system. According to Rosen [10], the following five approaches form the basis of second law analysis.

7.1.1.1 EXERGY ANALYSIS

Exergy concept is based on the assumption that a system (electric charge, high pressure water, heat, etc.) which is at equilibrium with the environment does not have the capacity to cause any change. But any system which is not in equilibrium with the environment has the potential to do work. The total exergy E of a system comprises of physical exergy E_{PH}, kinetic exergy E_{KN}, potential exergy E_{PT}, and chemical exergy E_{CH}.

1. *Physical exergy* results from the difference in internal energy, entropy and volume of the system and is measured with respect to the reference environment. It is used for analysis of processes, undergoing physical changes. The physical exergy is significant when the temperature or pressure of the substance deviates from its surroundings.
2. *Chemical exergy* is computed using the difference in chemical composition of a substance compared to its components in the environment. It is used for analysis of processes in which the chemical elements remain unchanged. It relates to the calorific value of the substance. The chemical exergy is an important parameter for fuels.
3. *Kinetic exergy* of a substance is attributed to its velocity with respect to the environment.
4. *Potential exergy* of a substance is attributed to its location above the ground level.

The total exergy E of a system may thus be expressed as

$$E = E_{PH} + E_{KN} + E_{PT} + E_{CH} \tag{7.3}$$

For simplification, only the thermal exergy (is the sum of physical exergy and chemical exergy) is taken into account for some engineering analysis. The effects of nuclear, magnetic, electrical, and surface tension are neglected in the analysis. The sum of the kinetic, potential, and physical exergies is referred to as thermomechanical exergy. The exergy analysis of a system deals with the evaluation of exergy inputs, exergy outputs, consumption of exergy, and accumulations of exergy. The exergy balance for a system is given by Rosen [10] as

$$\text{Exergy input} - \text{Exergy output} - \text{Exergy consumption} = \text{Exergy accumulation} \tag{7.4}$$

The exergy balance for a system can be written in various forms depending on the system under consideration.

Exergy balance for a system where exergy flows in and out of the system may be expressed as

$$\text{Rate of exergy change} = \text{Rate of exergy transfer} - \text{Rate of exergy destruction} \tag{7.5}$$

For the steady-state operation of a system, it is considered that the rate of exergy change is zero. The exergy balance equation in steady state reveals that the rate at which exergy is transferred to the control volume is greater than that transferred out of it and the resulting difference is the rate of exergy destruction within the control volume due to irreversibility. This analysis is very complex but provides an extra insight into the system. Thus, simplified forms of exergy analysis are often employed.

7.1.1.2 PHYSICAL EXERGY ANALYSIS

In order to simplify the exergy analysis for a system, physical exergy analysis is employed which includes all the components of exergy analysis except chemical exergy. Thus, this analysis has a limitation from its definition that it cannot be applied to processes in which the composition of stream changes (e.g., combustion). The physical exergy balance for a system is expressed by Rosen [10] as

$$\text{Physical exergy}_{\text{input}} - \text{Physical exergy}_{\text{output}} - \text{Physical exergy}_{\text{consumption}} = \text{Physical exergy}_{\text{accumulation}} \tag{7.6}$$

7.1.1.3 EXERGY CONSUMPTION ANALYSIS

The exergy consumption analysis does not include all the terms of exergy analysis. The exergy balance may be written as [10]:

$$\text{Work}_{\text{input}} - \text{Work}_{\text{output}} - \text{Exergy consumption} = \text{Work potential}_{\text{accumulation}}. \quad (7.7)$$

7.1.1.4 NEGENTROPY ANALYSIS

The term "negentropy" or "negative entropy" was used by Schrödinger in his book, *What is life?* Negentropy is regarded as a valuable commodity. The consumption of negentropy is equal to entropy production. A negentropy balance for a system can be expressed as [10]:

$$\text{Negentropy}_{\text{input}} - \text{Negentropy}_{\text{output}} - \text{Negentropy}_{\text{consumption}} = \text{Negentropy}_{\text{accumulation}}. \quad (7.8)$$

7.1.1.5 ENTROPY ANALYSIS

Entropy analysis of a system is based on the evaluation of the entropy inputs, entropy output, entropy accumulation, and entropy generation in the system. The balance of entropy for a system may be written as [10]:

Rate of change of entropy content of a body = Net flow of entropy in and out of the body + Creation of entropy inside the body (7.9)

This analysis deals with the losses occurring in the system due to irreversibility. Entropy generation in the system is proportional to the lost available work. It is identical to negentropy consumption and proportional to exergy consumption. Minimization of entropy generation means minimization of thermodynamic irreversibility in a system. This leads to increasing the exergy fraction of the energy. Thus, the method of entropy generation minimization is used in thermodynamic optimization of real systems where the irreversibilities are due to heat transfer, fluid flow and mass transfer. The entropy generation minimization method is distinct from the exergy analysis. The exergy analysis of a system is based on the first and second law of thermodynamics and the physical and chemical properties of the system w.r.t. the environment. On the other hand, entropy generation minimization method is concerned with minimizing the irreversibility that requires the relations between heat transfer and temperature differences. It utilizes the relation between pressure differences and mass flow rates. For optimization, the physical characteristics of the system are also considered.

In all the above discussed analysis, time dependence of the variables is not considered. Many other thermodynamic analyses are applied to system to optimize system performance. Finite time thermodynamics approach was introduced in applied physics as a new thermodynamic field with the objective of maximizing exergy delivery from a system over a finite time period [11, 12].

In this chapter, energy and exergy analysis of Scheffler dish collector is presented. The collector is installed at Om Shanti Retreat, Gurugram. It consists of two Scheffler dish concentrators which tracks the sun and concentrates solar radiation on the absorber surface, from top and bottom as shown in Figure 7.1. The incident solar radiation is absorbed by the absorber and transferred to the HTF (water) flowing inside the absorber. The water changes its phase and is stored as steam in the storage tank. There are heat losses from the absorber to the surrounding from both sides by convection and radiation. The collector tracks the sun throughout the day. Availability of solar radiation above threshold level is necessary for proper functioning, and hence, a good estimate of solar radiation would help in estimating the performance of such systems [5, 13]. Heat losses from collector are due to convection [14] and radiation from absorber surface [6, 15].

FIGURE 7.1 Scheffler dish concentrator system.

Convection and radiative heat losses have been studied by researchers in the past [16–19] by assuming the absorber temperature to be equal to mean fluid temperature or inlet fluid temperature. A major problem identified during the experimental analysis of this type of system at Omshanti Retreat Centre, Gurugram, was recording fluid temperature at inlet and outlet as well as the mass flow rate of the HTF.

Keeping in mind the difficulties faced during experimentation, in the present study, an attempt has been made to derive expression for overall heat loss coefficient for the collector using incident solar radiation, ambient temperature, wind speed, and absorber temperature as input which can be easily measured at any instant without much physical interventions with the system, and consequently derive the thermal efficiency of the collector. Experimental data is used to validate the model.

7.2 THERMAL ANALYSIS OF THE COLLECTOR

A steady-state energy balance for the absorber is given by

$$Q_u = [S - A_{abs} U_L \ (T_p - T_a \ )] \tag{7.10}$$

where S is the absorbed solar radiation measured in W, U_L represents the overall heat loss coefficient which includes the heat losses due to convection and radiation (W m^{-2} K^{-1}), A_a is the aperture area (m^2), T_p is the mean absorber temperature, and T_a is the ambient temperature (K).

Solar radiation absorbed by absorber can be computed using the following relation:

$$S = I_b r_b A_a \rho \gamma_1 \alpha_b + I_b r_b A_{abs} \alpha_b \tag{7.11}$$

where A_{abs} is the absorber area (m^2),$\rho\gamma_1\alpha_b$ is the product of reflectivity of the concentrator, intercept factor, and absorptivity of absorber, and I_b is the incident solar radiation (W m^{-2}). The first two terms represent the radiation reflected by the top and the bottom concentrators and finally absorbed by the absorber. The third term denotes the radiation absorbed by the absorber that directly receives it.

Radiative heat transfer coefficient (h_r) is computed using the following relation:

$$h_r = \sigma\varepsilon(T_p^2 + T_a^2)(T_p + T_a) \tag{7.12}$$

where $\sigma = 5.67 \times \frac{10^{-8} W}{m^2 K^4}$ and ε is the emissivity of the absorber surface.

Air flows over absorber surface which is spherical in shape and is insulated at the edge. The convective heat transfer coefficient for natural flow of air over the absorber is computed using the following relation [20]:

$$Nu_D = \frac{hD}{k} = 2 + \frac{0.589 Ra_D^{0.25}}{\left[1+\left(\frac{0.43}{Pr}\right)^{\frac{9}{16}}\right]^{\frac{4}{8}}} \tag{7.13}$$

Equation 7.4 is valid for Rayleigh number, $Ra_D < 10^{11}$ and Prandtl number, $pr \geq 0.7$, and the fluid properties are considered at $T_f=(T_p+T_a)/2$.

The overall heat loss coefficient:

$$U_L = h_c + h_r. \tag{7.14}$$

Thermal efficiency of the collector can be computed using the following relation using Equations 7.10–7.14 as

$$\eta = \frac{Q_u}{2A_{aI_b}r_b} \tag{7.15}$$

The exergy of the collector is a measure of the quality of energy in the system. Exergy output from the collector can be expressed in terms of the irreversibility resulting from an increase in temperature gradient and losses in pressure due to fluid friction as [21]:

$$Ex_{\text{out}} = \dot{m} c_p \left[T_{\text{out}} - T_a - T_a \text{In}\frac{T_{\text{out}}}{T_a}\right] + \frac{\dot{m}\Delta P}{\rho} \tag{7.16}$$

Exergy flows into the collector with the incident solar radiation. Exergy input to the collector can be expressed as

$$Ex_{in} = A_c I_b r_b \left[1 - \frac{4}{3}\frac{T_a}{T_s} + \frac{1}{3}\left(\frac{T_a}{T_s}\right)^4\right] \tag{7.17}$$

where T_s is the sun's temperature which is equal to 6000 K. The exergetic efficiency of the collector is given by

$$\eta_{ex} = \frac{Ex_{out}}{Ex_{in}} \tag{7.18}$$

7.3 RESULTS AND DISCUSSIONS

The solar thermal collector under the study is installed at Omshanti Retreat Centre located in Gurugram (28.2954° N, 76.8488° E). The specifications of the collector are shown in Table 7.1. For computation of thermal efficiency, data of solar radiation incident on the collectors, temperature of absorber surface, ambient temperature and wind speed were used. Mass flow rate, inlet and outlet temperatures of the fluid were also recorded to validate the result. These were recorded at 5 min interval from 10:00 am–12:20 pm in the month of April.

TABLE 7.1 Specifications of the Collectors

Operating Pressure, kg/cm^2	3
Absorber area, m^2	0.07
Aperture area, m^2	10
Absorptivity	0.95
Reflectivity	0.99
Intercept factor	0.99
Tracking	Single axis polar tracking

Figure 7.2 shows the variation of solar radiation and absorber temperature during the experimental period. The variation of the incident solar radiation was due to cloud cover on the day experiment. However, the system was operational on that day. This resulted in variation in the absorber surface temperature also. The absorber temperature is also affected by the convective and radiative heat losses which are dominant at higher surface temperature of the absorber.

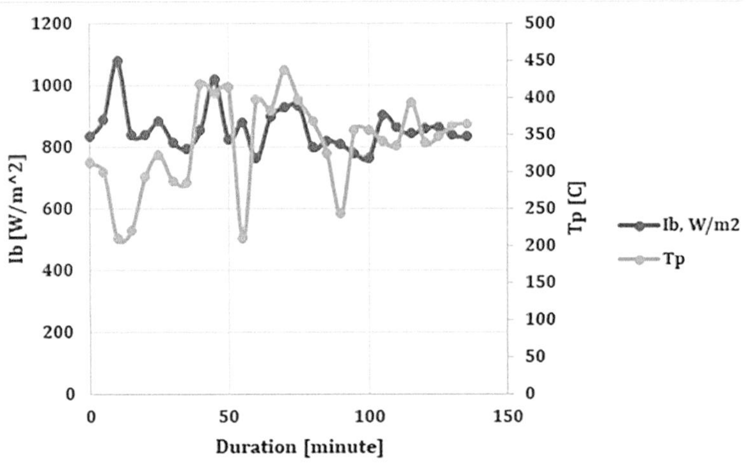

FIGURE 7.2 Variation of solar radiation and absorber temperature with respect to time from 10 am to 12.20 pm.

The efficiency of the collector is derived using the following relation:

$$Q\_u = \dot{m}\left[c_{pf}\left(T_b - T_{in}\right) + L_v + C_{pg}\left(T_{out} - T_b\right)\right] \tag{19}$$

where $\dot{m}$ is the mass flow rate (kg s^{-1}), C_{pf} and & c_{pg} are specific heat of water and steam respectively (kJ kg^{-1} K^{-1}), L_v is the latent heat of vaporization (kJ kg^{-1}), T_{in}, T_{out} and T_b are inlet outlet and boiling point temperature (K) of the fluid, respectively. Figure 7.3 shows the scatter plot of efficiency vs. $(T_p{-}T_a)/I_T$. The convective and radiative loss from the collector increases with the increase in the difference between the absorber temperature and ambient temperature. As a result, the efficiency of the collector decreases. Thus, a decreasing trend is observed in the graph.

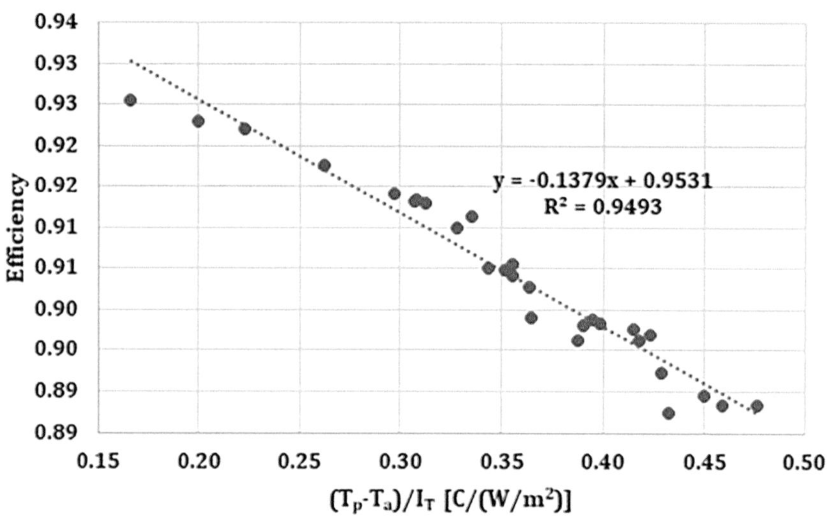

FIGURE 7.3 Efficiency vs. $(T_p - T_a)/I_T$.

To find the optical efficiency and overall loss coefficient, we fit a linear curve to this data.

Slope of the straight line $= - U_L/2C = -0.1379$ and intercept on vertical axis = optical efficiency = 0.9531.

Optical efficiency and U_L obtained using the proposed model described in previous section is compared with those obtained experimentally and is shown in Table 7.2. It is observed that the overall loss coefficient and optical efficiency of the collector obtained from both the methods are comparable.

The difference that is observed may be attributed to the variation in solar radiation and absorber temperature during the experimentation.

TABLE 7.2 Values of Overall Heat Loss Coefficient U_L and Optical Efficiency of the Collector

Parameters	Proposed Model	Experimentally
U_L, W/m^2 K	25.57	39.4
Optical efficiency	0.9344	0.9531

Energy and exergy efficiency of the collector was computed for the day. It is observed that energy efficiency varied from 89% to 91% and exergy efficiency varied from 48% to 67% as shown in Figure 7.4. It has been observed that exergy losses due to temperature gradient increases with the increase in surface temperature of the absorber. On the other hand, exergy losses due to fluid friction initially decreases with the increase in temperature, attains its minimum value, and then increases due to turbulent motion of fluid causing fluid mixing.

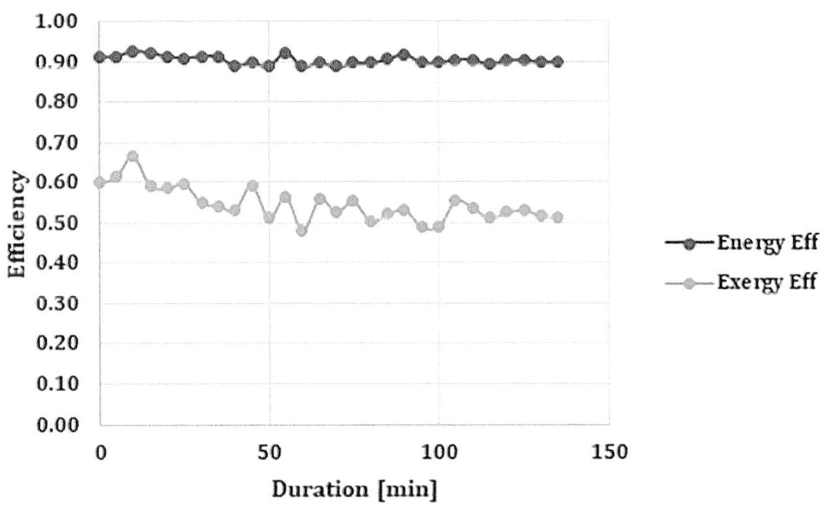

FIGURE 7.4 Energy and exergy efficiency of the collector w.r.t. time.

Average exergy efficiency of the collector over the entire experimental period was observed to be equal to 55% with a standard deviation of ±3.34% as shown in Figure 7.5. This suggests that the exergy output from the collector is almost uniform during the experimental period. This also suggests that these collectors can be used for industrial applications.

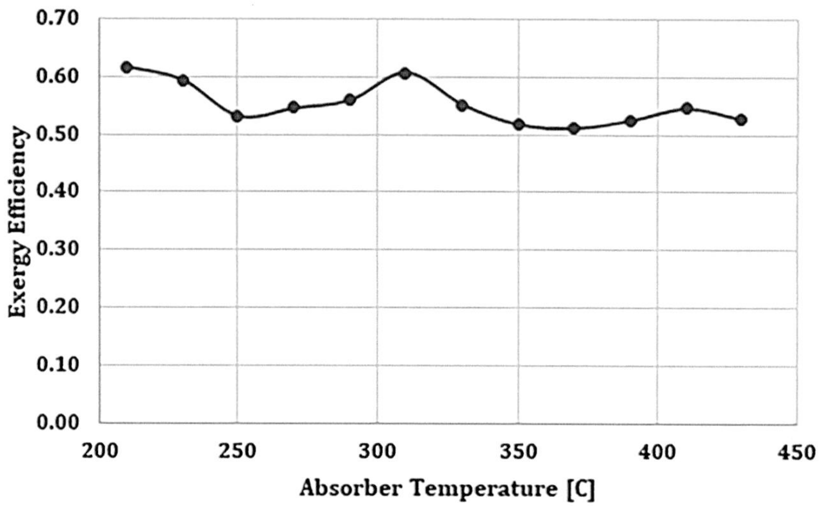

FIGURE 7.5 Exergy efficiency of the collector with respect to the absorber surface temperature.

7.4 CONCLUSIONS

In this study, the authors developed a mathematical model to find the thermal efficiency of a Scheffler dish collector consisting of two concentrators reflecting incident solar radiation to one absorber using only four input parameters, namely, incident solar radiation, ambient temperature, wind speed, and absorber surface temperature. These parameters were selected for the study as these can be recorded without any physical intervention of the collector. The following things have been observed while comparing the experimental and analytical results:

1. A difference of 1.9×10^{-2} has been observed for optical efficiency of the collector and 13.8 W m^{-2} K^{-1} is observed for the overall heat loss coefficient.
2. The difference in the result may be attributed to the variation in solar radiation over the experimental period which may resulted in temperature gradients on the absorber surface which otherwise has been considered as uniform.

3. This suggests that the proposed model can be used for thermal analysis of these types of collectors.
4. The model is simple and requires data which can be easily obtained.

Average exergy efficiency of the collector was found to be 55% with very small variation over the experimental period. Thus, these collectors are useful for industrial heat application.

KEYWORDS

- **Scheffler dish collector**
- **clean energy system**
- **exergy analysis**
- **experimental analysis**

REFERENCES

1. Duffie, J.A., Beckman, W.A. *Solar Engineering of Thermal Processes Solar Engineering*. 1st edn., John Wiley and Sons, Inc.: Hoboken, New Jersey, 1980.
2. Lamba M., Das S. Designing of parabolic trough collector using plane mirrors. *Journal of Physics: Conference Series* 2022, 2178, 012015. doi:10.1088/1742-6596/2178/1/012015.
3. Xie W.T., Dai Y.J., Wang R.Z., Sumathy K. Concentrated solar energy applications using Fresnel lenses: A review. *Renewable and Sustainable Energy Reviews* 2011, 15, 2588–2606.
4. Das S., Solomon S.S, Saini A. Thermal analysis of paraboloid dish type solar cooker. *IOP Journal of Physics: Conference Series* 2019, 1276(1), 012055. doi:10.1088/1742-6596/1276/1/012055.
5. Das S. Solar thermal power generation: Application of internet of things for effective control & management, in: *Handbook of Sustainable Development through Green Engineering and Technology*, CRC Press, Taylors & Francis Group, 2022, pp. 309–330.
6. Kanyowa T., Nyakujara G.V., Ndala E., Das S. Performance analysis of scheffler dish type solar thermal cooking system cooking 6000 meals per day. *Solar Energy Journal* 2021, 218, 563–570.
7. Gaggioli, R.A. *Thermodynamics: Second Law Analysis. ACS Symposium Series 122*. American Chemical Society: Washington, D.C., 1980: pp. 3–13.

8. Fuchs, H.U. *The Dynamics of Heat*. Springer-Verlag: USA, 1996.
9. Haouam A., Derbal C., Mzad H. Thermal performance of a gas turbine based on an exergy analysis. *E3S Web of Conferences* 2019, 128, 01027. doi:10.1051/e3sconf/201912801027.
10. Rosen, M.A. Second-law of analysis: Approach and implication. *International Journal of Energy Research* 1999, 23, 415–429.
11. Bejan, A. *Entropy Generation Minimization*. CRC Press: USA, 1996.
12. Mahanta, D.K. Second law analysis of solar flat plate collectors for optimum performance. *Thesis*. Department of Mechanical Engineering, Jadavpur University: Calcutta, India, 2001
13. Kumler, A., Xie, Y., Zhang, Y. A new approach for short-term solar radiation forecasting using the estimation of cloud fraction and cloud albedo. *Technical Report*. National Renewable Energy Laboratory, 2018.
14. Satpute, J., Chavan, K., Gulhane, N. Thermal performance investigation of concentrated solar collector using novel aluminum absorber. *Materials Today: Proceedings* 2018, 5, 4059–4065.
15. Barbosa, FV., Afonso, J.L., Rodrigues, F.B., Teixeira, J.C.F. Development of a solar concentrator with tracking system. *Mechanical Sciences* 2016, 7, 233–245.
16. Nene, A., Suyambazhahan, S., Ramchandra, S. Comparative analysis of performance of two scheffler solar concentrators having different concentration ratios. *International Journal of Engineering and Technology* 2017, *9*(2), 704–709.
17. Jagnade, S., Mishra, L., Baranwal, A.K., Chhabra, R.P. Laminar free convection in power-law fluids in a right-angle triangular duct with heated base. *Heat Transfer Engineering* 2019, 40(19), 1–26.
18. Talebizadeh, M.A., Javaran, E.J., Rahnama, M. Design, manufacturing and performance evaluation of a fixed focus Scheffler concentrator in Kerman. *Amirkabir Journal of Mechanical Engineering* 2018, 50(6), 419–422.
19. Das, S. Simulation of optimal exergy efficiency of solar flat plate collector. *Jordan Journal of Mechanical and Industrial Engineering* 2016, 10(1(, 51–65.
20. Jafarpur, K., Yovanovich, M.M. Laminar free convection heat transfer from isothermal spheres: A new analytical method. *International Journal of Heat and Mass Transfer* 1992, 35(9), 2195–2201.
21. Ngo L.C. Exergetic analysis and optimisation of a parabolic dish collector for low power application; 2012, Available on https://crses.sun.ac.za/files/services/conferences/annual-student-symposium-2012/22_November/6_Ngo.pdf.

CHAPTER 8

Energy and Medical Applications of Gold Nanoshells: A Review

A. S. DUTTA*

Department of Plastic and Polymer Engineering, Maharashtra Institute of Technology, Chhatrapati Sambhaji Nagar, Maharashtra, India

*Corresponding author

ABSTRACT

Nanoshells are basically the top coatings on cores comprising of different materials. They are easily synthesized by using one or two-step methods. The properties of nanoshells can be changed merely by changing the core-to-shell ratio. The shape, size, and morphological properties of these nanoshells can be altered easily. The properties of the coated nanoshells and the core material are different. Nanoshells as and when required can be synthesized with customized properties. Their demand is increasing in today`s world as they are being used in various fields for various applications.

8.1 INTRODUCTION

A nanoshell is a quasiparticle, which is having oscillating charge density. So it is also known as nanoshell plasmon. Nanoshell is made up of spherical-shaped nanomaterials and normally its core is having good dielectric properties. The most commonly used nanoshell is gold because of its inert nature [3]. The phenomenon of the oscillation of the ions of nanoshells is called plasmon hybridization. The tuning of hybridization depends on the

Advances in Energy Materials: Materials Fabrication, Properties and Applications.
Lishin T, Maciej Jaroszewski, and Sabu Thomas (Eds.)

DOI: 10.1201/9781779643674-8

oscillations of the ions of the inner and outer shells with varying energy levels. If the ions have low energy they respond strongly to the incident light. On the other hand, if the ions possess higher energy their response to the incident light is very weak [4, 5]. The most common method for synthesizing nanoshells is microfluidic composite foams. Nanoshells can be synthesized from any material, namely, semiconductors, insulators, or metals [1]. There are two types of nanoshells: oxide nanoshells (e.g., hollow silica nanoshells) and metallic nanoshells (e.g., gold and silver nanoshells) [2]. Nanoshells are gaining much importance in medical field and specially in the treatment of cancer. More fields of application of nanoshells are being explored and in future it is expected that the demand for nanoshells will be very high.

8.2 SYNTHESIS OF NANOSHELLS

The most commonly used nanoshell is gold (Au). The methods that are employed for synthesizing it are: (1) surfactant-assisted seeding method, (2) single-step deposition–precipitation (DP) seeding method, (3) sonochemical gold seeding method, (4) sandwiched gold seeded shell synthesis, and (5) one-pot synthesis. All these methods are used for silica core. Several different techniques are used when the core is changed. When the core material is polymer, the methods used for producing gold nanoshells are: (1) Solvent-assisted approach, (2) gold colloid seeding, (3) combined swelling-heteroaggregation, and (4) gold ion seeding method. It is also possible to synthesize hollow gold nanoshells by different methods. These methods are: (1) sacrificial template method, (2) template galvanic replacement of silver method, and (3) electrochemical synthesis of hollow gold nanoshells [6]. Nanoshells of gold coating and core material gold sulfide were first synthesized in 1994 [7]. Further modifications were made by changing the core to silica by Hala's group [8]. For binding silica core groups to the gold nanoparticles, the bridging units are amine groups. After this, it is mixed with gold salt solution. Then reducing agent is added for helping in the complete growth of the nanoshell. In recent days, trials have been taken by changing the core materials. Most trials are taken with different core materials and gold as the nanoshell. As the core material is changed, the properties also change in tune with the properties of the core material. We will discuss about each method of gold nanoshell synthesis in forthcoming section. As mentioned above, as the core material changes the synthesis method also changes. First, we will have an overview of the methods used when the core material is silica.

1. **Surfactant-Assisted Seeding Method**

   In the first step, silica nanoparticle is synthesized and 3-aminopropyltriethoxysilane (APTES) that is used as a linker is condensed on its surface. This linker provides the required amine groups to the gold molecules. In the next step, isolation of functionalized organosilane silica nanoparticle is carried out. This is followed by addition of gold colloidal solution to it. When the gold nanoparticle seeds get bounded to silica nanoparticle surface, the excess seeds are removed. For proper seeding of gold on silica two methods are being used. In the first method, an aged mixture of potassium carbonate and chloroauric acid is taken, and gold is seeded on required site by means of reduction reaction using formaldehyde. In second method, sodium borohydride is used for reducing chloroauric acid [7]. For providing a strong bonding between core molecules of silica and the gold seeds capping agents are used. One such example is mercaptoundecanoic acid, which helps in forming a covalent bond between the amine group of the core and carboxylic acid functional groups on gold seeds [9,10].

2. **Single-Step DP Seeding Method**

   This method is widely used for preparing gold nanoshells, which are being used as catalyst in producing carbon monoxide at low temperatures. As compared to previous method, this is less time consuming because it does not require aging of gold seeds and therefore nearly a period of two weeks is reduced in this method. The deposition and precipitation of the gold seeds directly occurs on the silica nanoparticles. Monodispersed silica was prepared by this method and the linker used was APTES. The amine groups of the core silica molecule are then seeded with gold hydroxide nanoparticles. In this method, the first step is mixing of sodium hydroxide with chloroauric acid. This increases the pH and hydrolyses the acid. This reaction produces gold hydroxide solution of mild yellow color. In the next step, the amine-grafted silica is added and the resulting mixture is heated at 70°C for 30 min. The product formation is indicated by an orange brown colored solution which is formed after the reaction. Another way of preparing this product was by using basic gold hydroxide solution, which was prepared by using chloroauric acid and potassium chromate. This solution was kept at room temperature under constant stirring for overnight period of time. Then it was mixed with freshly prepared sodium borohydride in the ratio of 10:1 by volume and added to the silica gold combination. As gold molecules underwent reduction reaction it was reflected by color change. In this method, the color changes were green, purple, or red. Gold seeds formation depends upon reaction temperature, time, and the amine groups surface modification. If unmodified amine groups are taken up, the resulting product shows variation in size (Figure 8.1).

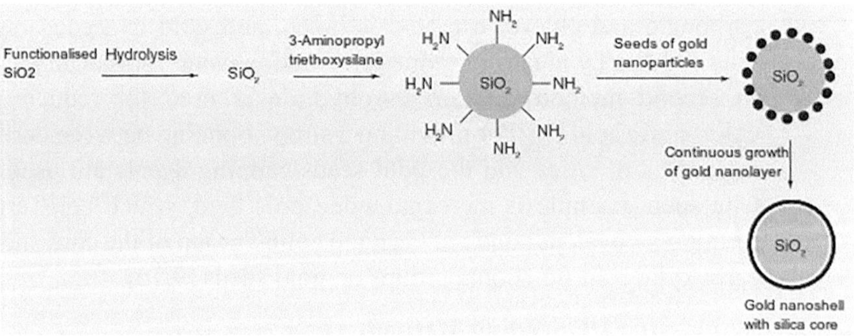

FIGURE 8.1 Gold nanoshell deposited on silica core.

Source: Adapted from Ref. [90]

3. Gold-Seeding Method by Sonochemical Process

This is an ecofriendly method in which silica spheres are coated with gold nanoshells by means of an ultrasound mechanism [11]. In this method, the preparation of gold seeds is not needed. It was first experimented by Calderon–Moreno's group. Normally in a sonochemical reaction, presence of ultrasound induces chemical changes, such as the formation, growth, and collapse of the bubbles of a liquid [12]. On similar pathway, silica core was coated with gold nanoparticles. The sonochemical reaction is as follows:

$$H_2O \rightarrow H^{\bullet} + {}^{\bullet}OH \tag{8.1}$$

$$AuCl^{4-} + 3H^{\bullet} \rightarrow Au^{(0)} + H^{+} + 4Cl^{-} \tag{8.2}$$

In this experiment, the researchers used 200 mg of silica submicrospheres which were mixed with chloroauric acid solution. This mixture was purged with argon for 2 hours to remove the dissolved gases. After purging, sonication was carried out. The temperature of this sonication reaction was in the range of 20–25°C. The last step is crystallization for which heating was carried out for 3 h at 500°C.

4. **Synthesis of Sandwiched Gold Seeded Shell**

   Gold nanoshells produced by this method have much enhanced properties as compared to previous methods [13]. In this method after seeding gold nanoparticles on silica core, another layer of silica nanoparticle is provided and therefore the name is sandwiched gold seeded shell. For forming the second layer of silica surface protected etching is carried out with help of chemicals, such as polyvinylpyrrolidone [14]. This etching helps in formation of porous silica shell. The porosity helps in diffusion of the gold seeds. Gold nanoshell is formed by slow addition of the gold salt. Here the silica core acts as a nanoreactor, which helps in anisotropical growth of the gold nanoparticles on its surface (Figure 8.2).

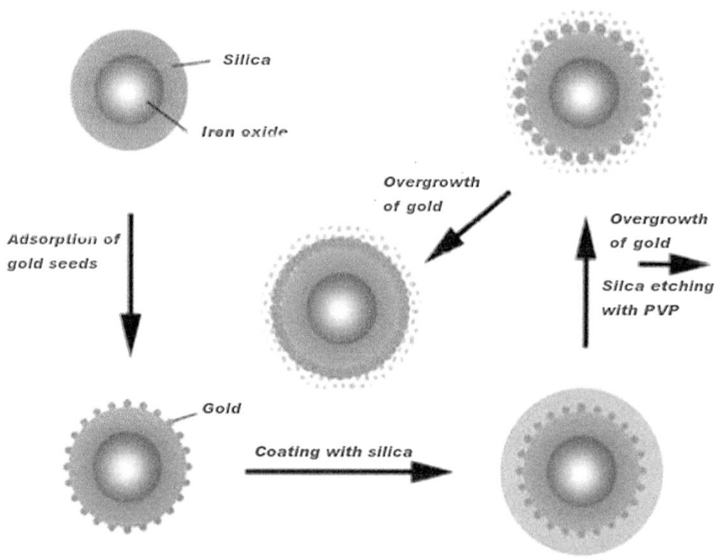

FIGURE 8.2 Gold nanoshell synthesized by sandwich method.

Source: Adapted from Ref. [6]

5. Gold Nanoshell Synthesis by One Pot Method

In the previous methods, nanoshell formation needed a suitable period of time which comprised of the various stages of the reaction. But every time such a lengthy process is not feasible and therefore researchers came up with this one-pot synthesis method [15]. As the name suggests here silica nanoparticles and gold nanoshells are formed sequentially in one pot. Here the reaction proceeds with nonfunctionalized silica and the dimensions of the gold nanoparticles are smaller as compared to the above methods.

The first step involves hydrolysis of APTES, which subsequently results in formation of amine functional groups on the silica molecule [16]. This reaction involves a stirring of nearly 30 minutes and addition of sodium citrate solution to the APTES mixture. For the next step that leads to gold nanoshell formation, additional sodium citrate is added and the mixture is heated to 100°C. This is followed by addition of chloroauric acid, which helps in reduction and nucleation of gold molecules. This helps in smooth bond formation between gold seeds and amine groups of silica molecule. For complete growth of gold nanoshell, the reaction is carried out for half an hour at 60°C [15].

We have discussed above five methods, which helps in covering a silica core with a gold nanoshell. When the core is polymeric in nature, the methods for synthesizing gold nanoshell change. Normally these methods are having multiple steps. These methods do not need the complex activation reaction for the functional groups present on the surface of the core molecules. The initial step is modification of the surface of the core (polymer), which is followed by seeding of gold nanoshell to its surface and ending up with the complete growth of gold on the surface.

1. Gold Nanoshell Synthesis in the Presence of Solvent

Without modifying the surface of the polymer group Ming`s group succeeded in developing a thin layer of gold metal on surface of core [17]. The core polymer taken was polystyrene, which was immersed in a mixture of gold salt and ethanol and acetone solvents. In presence of solvent, the polymer molecules swelled up and gold ions permeated in its core. It was found that this technique led to the formation of an uniform layer of gold nanoshell on the polystyrene core.

2. **Combined: Swelling and Heteroaggregation**

As above here also for swelling the polymer a solvent or a solvent mixture is used. The difference is that here gold ions reduction is not carried out and directly gold nanoparticles are used (Figure 8.3). After the polymer gets completely swollen, it is dried and solvent is removed and it causes shrinking of the polymer. This shrunk polymer core then traps the gold nanoshells on its surface [18].

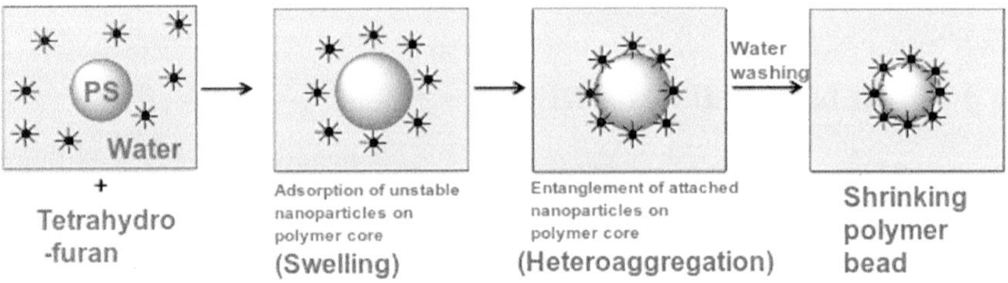

FIGURE 8.3 Gold nanoshell synthesized on polymer core.

Source: Adapted from Ref. [6]

3. **Gold Colloid Seeding Method**

This method is very widely used for seeding gold nanoparticles on polymer surface. Here the gold solution is in colloidal form. The bonding between colloidal gold seeds and the polymer surface is either by electrostatic interaction [19] or by covalent interaction [20]. As seen in above method, the most common polymer used here is polystyrene. Some ionic polymers are also being used in this method as the core molecule. Polyethylenimine (PEI) bearing a positive charge was attached by an electrostatic interaction to poly(2-methacryl(3-amide-2,4,6-triiodobenzoic acid)) (PMATIB), resulting in nanomaterial of PEI/PMATIB. This experiment was conducted by Yang`s group [21] and in the next step, they slowly added this mixture to a colloidal gold solution. Further the entire mixture was added in an aqueous solution of chloroauric acid and sodium citrate. It was observed that a thin gold layer was formed on the polymer surface.

4. **Gold Ion Seeding Method**

This method was studied by Klupp Taylor's group. Here they took a nonfunctionalized polymer as core molecule and lowered the concentration of the metal to be used as nanoshell which resulted in a partial growth of metal layer on the surface. Here also the polymer chosen was polystyrene and its surface modification was avoided. By using L-ascorbic acid the gold ions were reduced. Controlling various reaction parameters, such as temperature, pH resulted in different density and morphology of gold layer. The gold layer showed different morphological patterns like cup-shaped or dendrite shaped [22].

8.3 CORELSS GOLD NANOSHELLS

Such gold nanoshells are known as hollow gold nanoshells. In the above section, we have studied the different techniques by which gold nanoshells can be formed with two variations: one is the silica core and other is the polymer core. Now as the name suggests hollow shells would be studied. Following methods are used to produce such nanoshells: (1) replacing cobalt nanoparticles by sacrificial galvanizing, (2) replacing silver nanoparticles by templating galvanic replacement technique, (3) synthesizing gold nanoshell on silica core and later remove the core using hydrofluoric acid [23], (4) electrochemical build-up method, and (5) using a nonmetallic structure as core and gold as shell to prepare hollow gold nanoshells.

1. **Sacrificial Galvanizing Method**

   In this method, cobalt nanospheres act as reducing agent for synthesizing gold nanoshells. This method was first experimented by Liang's group [24]. In the presence of nitrogen gas, cobalt chloride aqueous solution $CoCl_2$ (aq) was reduced by using aqueous sodium borohydride $NaBH_4$ (aq). This step led to the formation of cobalt nanoparticles. In the next step, unreacted sodium borohydride was hydrolyzed to liberate hydrogen gas H_2 (g) [25]. Removal of sodium borohydride was essential as it led to the formation of wider diameter of the core. After the excess of sodium borohydride was consumed, the cobalt nanospheres were added to degassed chloroauric acid solution $HAuCl_4$ (aq). The standard reduction potentials of $AuCl_4^-/Au$ is 0.994 V and that of Co^{2+}/Co is −0.277 V, respectively. Due to this remarkable difference in the reduction potentials, electron transfer occurs between cobalt and gold ions resulting in Au^0. The reactions of this method are given below [26–28]:

$$2NaBH_4 + CoCl_2 \rightarrow Co^{(0)} + H_2O + 2NaCl + B_2H_6 \quad (8.3)$$

$$2NaBH_4 + H2O \rightarrow B_2H_6 + H_2 + 2NaOH \quad (8.4)$$

$$3Co + 2 \rightarrow 3Co^{2+} + 2Au^{(0)} + 8Cl^- \quad (8.5)$$

The structure of the gold nanoshell depends mainly on the diameter of cobalt nanospheres because the movement of gold molecules is restricted to the outer surface of this sacrificial cobalt molecule. Therefore the rate-determining step for this reaction is the synthesis of cobalt nanospheres. After the formation of gold nanohells in the last step, the reaction kettle is kept for oxidation in open air. This step of oxidation helps in oxidizing the remaining cobalt and finally leads to the formation of gold nanoshell with hollow core.

2. **Galvanic Replacement Method**

Similar to the above method, another element being widely used for synthesizing hollow gold nanoshell is silver [29]. The basic principle here is of replacement of silver with gold based on the differences of their redox potentials. In this method, electron transfer occurs between silver and gold ions to give Au^0·r The reaction occurring between $Ag^{(0)}$ and Au^{3+} is given below:

$$3Ag + AuCl_4^- \rightarrow 3Ag^+ + Au^0 + 4Cl \quad (8.6)$$

This experiment was first carried out by Xia`s group [30, 31] who took trials for synthesizing gold nanoshells by using silver nanoparticles of different morphologies. It was concluded that morphology of gold nanoshell was dependent on the morphology of silver nanoparticle. A variation in this method was adapted by using nonaqueous solvents, such as toluene and chloroform. In this method, octadecylamine was used as a carrier and it facilitated transfer of silver nanoparticles into the solvent. These silver nanoparticles were further mixed with chloroauric acid. Presence of organic solvent helps to modify the surface of gold nanoshells easily. Based on this method, Drezek`s group encapsulated the inner part of gold nanoshell with ferric oxide. In this trial, first, the core which was of silver was covered with ferric oxide nanoparticles and this was covered again by a layer of silver nanoparticles by this sacrificial templating technique.

4. **Electrochemical Built-Up Method**

In the methods discussed above, synthesis of core templates is very important. Above methods also used certain capping and stabilizing agents so that the desired thickness of gold nanoshell is achieved. Also a very important step is the gold reduction in the final stage [32]. Proper core template synthesis is very important because if this is not monitored properly it causes aggregation, which leads to uncontrolled shell thickness. Usage of stabilizing and capping agents restricts the surface modification capability of the nanoshells as per the requirement.

For avoiding these complications, researchers have tried to prepare a clean and hollow gold nanoshell. The nanoshell was synthesized on indium tin oxide (ITO). On the surface of ITO silver nanoparticles were electrodeposited. This method does not involve the usage of any other reagents. The method does not require formation of templates [33]. The hollow gold nanoshells were formed by a galvanic replacement reaction between chloroauric acid and the silver nanoparticles. The surface molecules of these gold nanoshells could be easily modified as per the end-use application purpose [34].

5. **Hollow Gold Nanoshell Synthesis by Using Nonmetallic Structures**

As against hard metals that were used as templates in the above methods, in this method soft templates, such as emulsion droplets were used. Usage of nonmetallic templates is advantageous as compared to metallic templates because metallic templates prove harmful for biological applications. The soft templates also become compatible with any other material very easily.

a. **Nonmetallic Structures: Emulsion Droplets**

Synthesizing gold nanoshell in the presence of surfactant has been discussed above. For reacting with chloroauric acid, the various chemicals used were APTES, gelatin, polylatide, etc.

For stabilizing the nanoshell capping or stabilizing agents were also added. Further modifications in this method were incorporated. One such option was using mechanically stirred and stabilized emulsion method. This system was free of surfactants. The solvent used is toluene which forms the organic phase. For transferring gold ions to this organic phase, phase transfer agent, such as tetra-octylammonium bromide was used. The stirring is continued and sodium borohydride solution is added to this mixture. After this addition, nanoparticles of water containing sodium borohydride are formed in the organic phase. In the next step, gold ions are reduced to form hollow gold nanoshells. These shells are capped with dodecanethiol at the interface of organic and aqueous phase. Following this method, fluorescent dyes and other nanomaterials can be easily used as an encapsulating material.

2. **Nonmetallic Structures: Biomaterial Templates**

Biomaterials have proved to be good quality template material for synthesizing metallic nanoparticles [36, 37]. For example, viruses have the capability to access the interior of the material and extensive tailoring ability in terms of chemical properties. Also they have site specific heterogenous chemistry. Generally, viruses are protected with a layer of protein known as capsid. On the surface of the capsid various amino acids, such as cysteine and aspartic acid are present. These amino groups help in the anchoring of the gold nanoparticles. The thiol groups are compatible with the amino groups of the viruses [36, 37]. So with the help of these viruses metallic nanoparticles of various shapes can be easily synthesized. If hollow interior is desired then the removal of core genetic material is also easy. The process for synthesizing gold nanoshells by using viruses is carried out in two steps. In the first step, its stock solution is prepared and in the second step directly the gold nanoparticles are added. This leads to the formation of gold layer on the surface of the virus. The reducing agent used in this process is hydroxylamine hydrochloride [35].

8.4 SYNTHESIS OF OTHER TYPES OF GOLD NANOSHELLS

Gold nanoshells are also synthesized in different shapes, such as nanostars [43], nanoflowers, and also multilayered gold nanoshells. These varieties can be synthesized using different methods. Out of these, the multilayer gold nanoshells have multiple bands and here silica layer is placed as a spacer. First method of synthesizing them was by inserting gold nanospheres inside the gold nanoshell and silica layer was placed as a spacer between the layers. The multilayer gold nanoshells show good surface plasmon resonance (SPR) and this effect depends upon the ratio of diameters of silica spacer and the gold nanoshells. As it is multilayered, different layers can take up different functions, such as outer layer can act as optical condenser and focus light on the inner core layer of gold nanoshell.

In the first stage of synthesis, APTES is added to the gold nanoparticles solution for enhancing anchorage of amine group to gold particles as well as for formation of the thin layer of silica spacer. The thickness of this spacer layer normally ranges between 4 and 10 nm. The next step is condensation reaction in presence of sodium [38]. Tetraethyl orthosilicate ammonium hydroxide are used normally in lower concentrations for producing smaller size of the gold nanoparticles [39]. They also facilitate the proper growth of the silica layer which helps to control the charge destabilization. APTES with help of its amine groups helps in forming colloidal gold layer which acts as the precursor [40]. For the complete growth of the gold nanoparticles reduction of chloroauric acid is carried out. By adapting the Stober method, proper seeding of silica is carried out which helps in formation of a silica layer with a precised thickness. The excess of silica is etched out by means of a hydrolysis reaction.

A variation in the above method was carried out by synthesizing silica layer in presence of thiol groups instead of amine groups. Here mercaptopropyl triethoxysilane was coated on the core and its thiol groups acted as an anchor for the gold nanoparticles. Further it was followed by reduction of chloroauric acid, which led to formation of gold nanoshell in one pot (method has been already discussed above) [41]. So in this variation, the formation of precursor colloidal gold is not required. This method is advantageous as many intermittent steps, such as surface modification, phase transfer, and repetitive washing are avoided here. By adapting the sacrificial template method which has also been discussed above, empty spacer can be synthesized instead of silica spacer by coating silver on the gold nanoparticles [42].

The gold nanoparticles that are synthesized with spikes or protruding arms have the capability to develop a powerful electromagnetic field at their tips [44]. Also they are of anisotropic nature and therefore they show different properties when analyzed from different directions. This makes them suitable for being used in surface-enhanced Raman spectroscopy, in nonlinear optical applications [46] and also for detection of the biomolecules [45]. The spiky gold nanoparticles have good photothermal heating capacity. Because of this property, they are being used in diagnostic and therapeutic applications. The PR effect of these nanoparticles can be varied by controlling their tip length.

They can be synthesized by means of seeding or by using structure directing agents for incorporating nanoparticles on self-assembled nanostructures of block copolymers or polymeric beads. One such example is of cadmium–selenium nanoparticles which were coated with a mixture of iron oxide and zinc sulfide [47]. These were then incorporated on the soft gold nanoparticles which finally grew unto spiky gold nanoshells. The first step of this reaction is reduction of silver diamine cation to silver nanoparticles by using sodium borohydride. Through this reaction silver ions got seeded on the core surface. This was followed by overnight aging to assure entire consumption of sodium borohydride. This seeded core was mixed with a growth solution containing partially reduced gold ions. This growth solution was a mixture of cetrimonium bromide, silver nitrate, ascorbic acid, and chloroauric acid [48]. Reduction rate of gold ions and its shape is decided by the concentration of the silver ions.

For controlling the topography of these nanoshells various surfactants were used as the surface directing agents. Polyvinyl pyrrolidone was used as a structure directing agent by Ling`s group for increasing spike length from 50 to 130 nm [49]. But these agents get strongly absorbed on the surface and pose great difficulty while removing them. Branched gold nanoshells were synthesized by using seeding method devoid of surfactants. This was first experimented by Mosquera`s group [50]. In this method, the core was covered with chitosan and was used for binding the gold seeds which were capped with citrate molecules through electrostatic interaction. The growth solution contained chloroauric acid and potassium carbonate. Successful trials were also taken by using glycol chitosan against pure chitosan. The surfactant-free method is a clean process and can be easily functionalized when required. The factors controlling the different degrees of anisotropy as well as degrees of optical responses were the ratio of chloroauric acid and potassium carbonate, ascorbic acid and the nanoparticle seed precursor [51].

8.5 PHYSICAL PROPERTIES OF GOLD NANOSHELLS

Mostly gold nanoshells are used for tumor treatments and the desired properties for this application are localized surface plasmon resonance (LSPR) and radioactivity Scientists have made use of these excellent properties for various purposes, such as noninvasive detection and *in vitro* diagnostics. The radioactivity property of these nanoshells is very well used for radiotherapy and radionuclide imaging. Gold has high atomic number of 79 which is useful for radiotherapy treatment purposes.

8.5.1 LSPR OF GOLD NANOSHELLS

When light is incident on a nobel metal, its conducting electrons get stimulated and show collective oscillation. This phenomenon is known as "plasmon." The plasmon resonance (PR) is seen when the frequency of the incident photon is in resonance with the frequency of the oscillating electrons [53]. Mostly this resonance is observed on metal surface and shows a rapid declination with the depth of the metal. As phenomenon is observed on the surface only it is called "surface plasmon resonance (SPR)" [54]. This phenomenon gets a localized effect when it is restricted to smaller volumes and sizes of material, for example, nanoparticles and is named as LSPR. This phenomenon induces two effects in gold nanoshells, namely, extinction coefficient and the enhancement effect of localized electromagnetic field. Here extinction coefficient refers to the intensity of light absorption in a particular medium [52].

8.5.2 RADIOACTIVITY OF GOLD NANOSHELLS

Gold is showing nuclear properties and is a radionuclide. It has two isotopes ^{198}Au and ^{199}Au which are commonly used in biomedicines. ^{198}Au emits the beta particles with an energy of 0.96 MeV [57]. The energy emitted while radiating gamma particles is 412 keV [57]. The content of radioactive atoms in ^{198}Au nanoparticles is very high and therefore a small quantity of it is suitable for radioactive based treatments. The high energy emitted when beta particles are emitted are useful for treatment of tumors whereas the gamma rays emission can be utilized for penetrating the soft tissues and in tomographic scanners.^{199}Au emits beta and gamma particles at a lower energy than the previous one. Here for beta particles energy is 0.45 MeV and for gamma rays it is 208 keV. This energy level makes it more preferable for single-photon emission computed tomography scanners [58]. For this reason, nanoparticles of ^{199}Au result in more target specific activity and thus facilitate the delivery of the drugs to the tumors [52].

8.5.3 ATOMIC NUMBER OF GOLD

The absorbancy of X-rays is directly proportional to the atomic number, specifically beyond 53. The atomic number of gold is 79 and therefore it shows very high tendency of absorbing the X-rays. This acts as a radiotherapy sensitizer in oncology treatment and minimizes the damage to normal tissues [55]. Gold molecules help to increase the cross section of a cancer cell and also have the capability to emit Auger, Compton, photo, and other secondary electrons. These secondary electrons have the capability to directly ionize DNA molecules by reacting with the water in tissues for liberating free radicals. These radicals then bind to DNA molecules and electron transfer to these DNA molecules takes place and finally they get oxidized. This is the procedure of radiotherapy sensitization. The enhancement effect of X-rays by means of the gold nanoparticles is termed as dose enhancement factor [56]. This factor relies on the size and concentration of the nanoparticles, different energies of the emitted X-rays, 3D dose distribution, etc. Trials were carried out on mice suffering with tumors by Hainfield and his group. Radiation carried out with the help of these gold nanoparticles proved to increase the survival time of the mice. Also gold nanoparticles because of the above advantage are used as an imaging agent for cancer diagnosis [52].

8.6 CHEMICAL PROPERTIES OF GOLD NANOSHELLS

8.6.1 COMPATIBILITY

Gold nanoshells have the ability to form stable bonds with other compounds, such as nitrogen and sulfur-containing groups. This compatibility aspect of the gold nanoshells makes it preferred material for ligand formation and also for some polymers suited for some specific applications. The stable structure and compatibility of gold nanoshells makes it suitable for being used in drug delivery applications [59].

8.6.2 CATALYTIC ACTIVITIES

Gold nanoshells have been successfully used as biomimetic catalysts. First trial was taken by Haruta et al. in 1987 for the oxidation reaction of carbon monoxide [60]. Gold nanoshells have been effectively used in enzymatic activity for enzymes, such as peroxidase (POD), glucose oxidase (GO), and catalase. Some of these enzymes have been used in tumor treatments. The POD-mimic activity of gold nanoshells was used for catalyzing hydrogen peroxide [62]. Similarly, the combination of GO and gold nanoshells has been used for oxidizing glucose and hydrogen peroxide under aerobic conditions [61].

8.6.3 BIOLOGICAL ACTIVITIES OF GOLD NANOSHELLS

Gold nanoshells or particles not only have the activation capability but also have intrinsic biological activity capacity. This makes it the most widely used material in tumor therapy. The activity of the nanoshells depends on their size, if they are less than 2 nm then they can easily interact with DNA molecules and kill the cancerous cells. But this is not seen if the particle size is larger than 2 nm. If the particle size is nearly 5 nm these nanoshells can change molecular conformations and inhibit the activity of the tumor. Also they have the capacity to inhibit cancer cells proliferation [63] as well as increase the content of reactive oxygen species in a cell.

8.7 APPLICATIONS OF GOLD NANOSHELLS

8.7.1 GOLD NANOSHELLS IN DRUG DELIVERY

Gold nanoparticles have been successfully used as drug-delivery vehicles because of the following advantages: (1) its large surface area which facilitates higher drug loading, (2) its ability to modify the surface according to the targeted ligands, (3) its enhanced permeability and retention effect, and (4) its nature of releasing drugs in controlled manner [66]. Tumor necrosis factor-alpha (TNF-α) is of toxic nature and this acts as hinderance in its usage for therapeutic applications [65]. When they were combined with polyethylene glycol-coated gold nanoparticles and used as a drug delivery system it was observed that the toxicity of TNF-α was reduced as well as this system efficiently destroyed the targeted tumor. For delaying tumor growth, thermally induced and pretreated gold nanoparticles were used. This drug delivery system was administered intravenously in the body and it was found to suppress blood flow of the tumor as well as destroying it completely. No accumulation of this combination was found in the liver, spleen, and other organs of the body. Further trials are being conducted to record its clinical efficacy as well as safety and pharmacokinetics aspects.

A chemotherapy drug methotrexate when conjugated with gold nanoparticles was found to suppress tumor growth in a mouse infected with lung cancer. Composites of nanoshells and hydrogels were successfully used as a drug-releasing system. This system was used for loading drugs, such as insulin, lysozyme, etc. Hollow gold nanoshell was used for encapsulating the active enzymes, such as horseradish POD. Trials using nanoparticles of gold has been conducted but predicting a specific drug for combining with them to obtain the desired results is not yet achieved. The physical dimensions of the nanoparticles determine their intracellular uptake capacity. The SPR peaks of the gold nanoshells lies in the near-infrared region. Because of the core-to-shell radius ratio nanoshells have limited applicability in this field. DNA modified nanoparticles have the ability to break the gold sulfur bond and release the thiolated DNA strands and this can utilized for controlled drug release applications. In future gold nanoparticles functionalized with polyethylene glycol chains might be most preferred system for treatment of cancerous tumors. As the conjugate compound formed is very big in size (diameter more than 50 nm) the drug is not released in a smooth manner. To overcome these issues, gold nanoparticles conjugated systems that can be specifically directed to a specific organ or site of disease should be found out [64, 66].

8.7.2 GOLD NANOSHELLS IN GENE DELIVERY

For the gene therapy, efficient carriers are needed for safe delivery of DNA and RNA molecules. Also the carrier molecule should have the capacity of rapid cellular uptake as well as protect DNA and RNA from getting degraded in the blood stream. For satisfying all these requirements, gold nanoparticles are the most opted material. Trials were taken by Braun et al. by coating small interfering RNA with gold nanoshells and using it for gene expression silencing in a group of cells [67]. It was observed that the release of small interfering RNA temporarily controlled the gene-silencing process. Similarly, the expression of oncoprotein of the glioma cells was restricted by using a complex formed of small interfering RNA and gold nanoparticles. This structure was named as spherical nucleic acid. These trials were conducted by Jensen et al. [68]. It was observed that the spherical nucleic acids efficiently neutralized the oncoprotein and no transfection or chemical modification was seen during this reaction. When spherical nucleic acids were injected in a mice model, it penetrated the blood–brain barrier and also reduced the volume of the tumor in mice without any side effects. For delivering small interfering RNA to the tumors a nanosystem comprising of gold nanostars and small interfering RNA was used for targeting heat shock protein 72 and hyaluronic acid of the cells [69]. This nanosystem inhibited the expression of heat shock protein 72 in triple negative breast cancer cells. This nanosystem showed minimum side effects [66].

8.7.3 GOLD NANOSHELLS IN ADJUVANT AND VACCINE DELIVERY

Various therapies are being put forth for building resistance in the body to fight cancer. In medical terms, this resistance is known as Immunotherapy. In recent years, the gold nanoparticles have emerged out as a successful carrier of the cancer vaccine because of their above-mentioned advantages. The vaccines delivered by means of the gold nanoparticles were capable of delivering antigens to stimulated cytotoxic T lymphocytes as well as to the dendritic cells. The cytokines present in the tumor cells of the body have a tendency to suppress the immunity power of the body, therefore, adjuvants are co-delivered along with the vaccines. Lee et al. took such a trial by using antigen vaccine red fluorescent protein and CpG oligodeoxynucleotide as the adjuvant vaccine along with 7 nm gold particles as a delivery mode [70, 71]. Trials were conducted on mice and remarkable antitumor activity was observed [66].

8.7.4 GOLD NANOSHELLS IN IMAGING APPLICATIONS

8.7.4.1 LIGHT SCATTERING IMAGING

The light scattering ability of the gold nanoparticles is nearly 1 million times higher than that emitted by fluorescent dye. The gold nanoparticles also show good photostability. Also the light scattered by them does not blink. Because of these attractive properties gold nanoparticles in various shapes, such as nanoshells, nanostars, etc., are used widely in optical imaging. A successful trial was conducted by Sokolov et al. by conjugating gold nanoparticles (approximately 12 nm in diameter) with the antibodies and using it against epidermal growth factor receptor (EGFR) cells [72]. These EGFR cells were prominent in malignancy growth and when exposed to the above conjugated compound they were individually visualized in the scattered light. It was also observed that these EGFR cells could be easily observed against the healthy cells by using simple dark field optical microscope. These observations were made by El-Sayed et al. [73]. On conjugating gold nanostars with hyaluronic acid the malignant cells could be easily observed using dark field optical microscope.

Another field of usage of the gold nanoparticles is the optical coherence tomography. This test provides the optical cross sectional images of the tissues depending on the intensity of the scattered light. Gobin et al. observed that when gold nanoshells were injected in the body the images of the malignant tumors were enhanced [66, 74].

8.7.4.2 PHOTOACOUSTIC IMAGING (PAI)

When a material is optically excited it expands and simultaneously acoustic waves are generated and the images captured during this phenomenon are the PAI. If a material is exposed to pulsed laser irradiation, then its temperature rises resulting in ultrasonic waves which are detected by the PAI method. Images obtained from PAI method have more clarity than those obtained from optical methods because the ultrasonic waves penetration capacity is more than the normal light. As gold nanoparticles have the ability to tune the optical properties and easily convert the photothermal energies, they are highly preferred for PAI methods. For treating neovasculature tumor, Nie and his group used functionalized gold nanostars in conjunction with cyclic arginine-glycine-aspartic acid peptides which showed high affinity for the receptor cells of the neovessels of the tumors [75]. The PAI images of the neovessels that were obtained showed high spatial resolution and also the depth of the imaging was quite deep. In another trial, the tumors were targeted with a conjugated compound of pH-low insertion peptides with gold nanostars [76]. After conjugation, the activation of the functionalized gold nanostars was done in a mild acidic environment of the tumors. When PAI images were taken after a day of injecting the conjugated compound it was found that the functionalized nanostars showed 50% higher tumor signal as compared to the nonfunctionalized nanostars [66].

8.7.4.3 X-RAY IMAGING

The 3D anatomical images taken for cancer infected persons/animals which is having very high resolution as well as economical is computed tomography (CT). In common terms, we hear about CT scan which is this X-ray-based scanning test. Contrast agents, such as iodine-based compounds are injected before scanning so as to obtain a clear distinctive image of infected versus healthy tissue. As the atomic number of gold is high [79], it shows strong X-ray attenuation coefficient. This helps in getting a contrast image between tumor tissues and the healthy tissues. Widely used gold nanoparticle-based contrast agent is the conjugated compound of gold nanoparticle with deoxyglucose [77]. Enhanced contrast images of the cancer cells were obtained when they were injected with this compound. Similar results were obtained when a compound of acetylated dendrimer and gold nanoparticles was used [78]. When polyethyelene glycol conjugated gold nanostar was injected in a cancer infected mice, the CT scan images obtained showed nearly 10 times enhancement as compared to normal images [79]. As the gold nanoparticles can be easily functionalized by using groups, such as ligands, peptides, antibodies, etc., so Hainfeld et al. used this property for treating breast cancer cells [80]. The conjugated compound was made up of gold nanoparticles, polyethyelene glycol, and Herceptin as the antibody. The CT attenuation of these tumors was enhanced by the usage of this compound. Kim et al. used prostate-specific membrane antigen for attaching on the surface of the gold nanoparticle and used them as prostate cancer targeting CT probes [66].

8.7.5 GOLD NANOSHELLS IN CANCER TREATMENT

Hyperthermia treatment is conducted at high temperature and is successful in killing many infected cells of the tissues. Also this therapy has increased the overall life expectancy when conducted in combination with chemotherapy and radiotherapy treatments. This therapy can be delivered to the body by various modes, such as externally, interstitially, or through tubes or ducts. The sources for generating heat are ultrasound, microwaves, or radiofrequency waves [82]. This therapy shows tumor selectivity and restricts the growth of the tumor cells by attacking its vascular system. But it lacks in certain aspects, such as deeply located tumors cannot be treated by this method and also it shows a lack of tumor specificity as well as temperature tolerance limit of the body after the initial treatment.

When laser producing nonionizing electromagnetic radiation is applied on metallic nanoparticles, such as gold, heat energy gets converted because of the electronic excitations of that metal. Also these laser radiations can be tuned to the SPR frequency of the nanoparticles. The most widely used nanoparticle for hyperthermia is the gold nanoshell. Normally these nanoshells contain silica as the core material. This combination helps to shift the resonance peak to the near infrared region wherein maximum blood and tissues become transmissive in nature [83]. Trials were conducted on cancer infected mice and it was found that when the laser radiation which was tuned to the nanoshell resonance peak, was applied it resulted in rise in the average tumor temperature by 37°C. All mice survived for three months without any danger of tumor reoccurance. The other group of mice who were not subjected to this therapy showed 9°C rise in the average tumor temperature and had to be killed within 20 days because of the uncontrolled tumor growth [84]. A combination of laser radiation and polyethylene glycol conjugated gold nanoshells was used for treating xenografts of the prostrate cancer in human beings. When a comparison was done on mice by giving them only laser radiation and a combination of gold nanoshells and laser radiation, it was observed that in later case tumor showed complete resolution in maximum mice (Figure 8.4). In the case of the mice treated with only laser radiation, it was found that the tumors remained unaffected [85]. The major limitation of this therapy is that the laser radiation used as a source has a limited range of penetrating capacity [81].

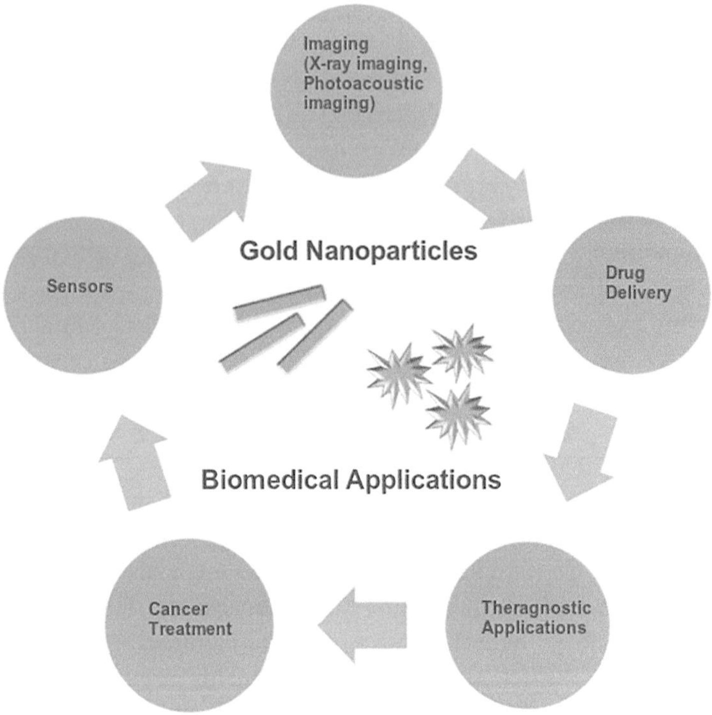

FIGURE 8.4 Applications of gold nanoshells.
Source: Adapted from Ref. [91]

8.7.6 GOLD NANOSHELLS AS RADIOSENSITIZERS

In treatment of cancer, radiation plays a vital role. Depending on the type of tumor radiation of various sources, such as gamma rays, photon-based radiations are used for the tumor treatment. One such radiation is used to damage the DNA associated with the tumor so as to restrict the tumor growth. This radiation is containing photons as the source. It acts by different means, such as by direct ionization or by producing free radicals [86]. The intensity of these rays is in range of 8–18 MeV. Gamma rays are used for generating the photons because they have the capability of low linear energy transfer and this helps in depositing a large fraction of the radiation in the healthy tissues before focusing on the infected tissues. This results in excitation of the core electrons of the atoms and release of the Auger electrons. As compared to the biological tissues the atomic number of gold is higher and it results in easy release of the Auger electrons from the gold nanoparticles. These Auger electrons are capable of attacking the damaged cells and their associated DNA. Hainfeld et al. have taken successful trials with gold nanoparticles having a diameter of nearly 2 nm for radiation purpose and got successful results for the same [87]. Trials were taken on mice suffering from breast cancer and their survival rate increased from 20% to 86% when they were subjected to gold nanoparticles sensitization. These nanoparticles have a disadvantage of short half-life as they are immediately excreted out from the kidney. For reducing the dosage of gold trials were taken with larger sized nanoparticles. One such trial was taken for ovarian cancer patients and the radiation consisted of the gold nanoparticles conjugated with thio-glucose molecules [88]. As the metabolism rate of cancerous cells is higher than healthy cells, the glucose coating resulted in an increase in the uptake of the cell by nearly 30%. Another trial was taken with 30 nm gold nanoparticle conjugated with an antibody, Herceptin [89]. The trails were conducted on the mice and it was observed that the radiations resulted in nearly 46% suppression of the tumors. When the tumors were radiated with only gamma rays the tumor suppression was only 16%. The size of the gold nanoparticles suitable for radiation therapy is not yet decided because some researchers are of the opinion that larger nanoparticles release more secondary electrons and consequently dosage delivery to cytoplasm gets reduced. While some researchers claimed that as size increases the accumulation capacity in targeted tumors increases [66].

Another type of radiation therapy is based on ions. Instead of using high energy gamma rays here ion beams are used for radiation. The ion beam has a tendency of linear energy transfer at the end point and this is termed as Bragg peak. The energy of the ion beam is directly proportional to the location of the Bragg peak. The mechanism suggested that for utilizing the gold nanoparticles for radio sensitivity purpose the excitation of the plasmons present on surface should be carried out, which will increase the yield of secondary electrons. Li et al. took trials of using gold nanoparticles for proton therapy. For this experiment, trials were taken with proton beams of varying intensities (10–25 keV μm^{-1}) and gold nanoparticles with sizes ranging from 5 to 10 nm. The radiation was effective with high energy proton beams. Similarly, with gold nanoparticles, as the size increased the effect was more intensified. Kim et al. took trials on tumor infected mice by using 14-nm-sized gold nanoparticles along with the proton beam. Twenty-four hours after the radiation, it was observed that the tumor regression was achieved. This was compared with trials taken on tumor infected mice and treated with only proton radiation. Here the tumor regression found was 37–62% [66].

8.7.7 GOLD NANOSHELLS FOR ENERGY GENERATION

8.7.7.1 HYDROGEN GENERATION

The gold nanoshells are effective materials which can be used for combating the changes in the climate. As per researches, it has been found that if the gold nanoparticles are coated with semiconductors they can produce hydrogen from water molecules and the efficiency is nearly four times the conventional processes. Basically, it is the reduction of water molecules to generate hydrogen atoms. The gold nanoparticles for this process are synthesized in star shape. In the actual process, the visible and infrared light was tapped and was absorbed by the gold nanoparticles. After this, electrons generation was fastened up due to the presence of gold nanoparticles and as a result the reaction speeded up.

The gold nanoparticles were coated with titanium dioxide and exposed to various wavelengths of light (particularly infrared) and it successfully generated hydrogen atoms at a very fast rate. These hydrogen atoms can be used up in absence of sunlight as a source of solar energy generator [91].

8.7.7.2 ELECTROCHEMICAL BATTERIES

For electrochemical energy storage, gold nanoparticles and carbon-based composites are widely preferred as they have the required surface chemical properties, efficient electron conductivity, and chemical stability. For supercapacitor graphene, gold nanocomposite has been successfully used and it was observed that it exhibited higher specific capacitance and excellent longevity. The capacitance retention capacity was very high as compared to the conventional electrodes. This electrode showed sensing for hydrogen peroxide up to the limit of 1 μm. The gold nanoparticles were synthesized by green synthesis method and the reducing agent used was Solanum nigrum leaf [92].

8.7.7.3 FUEL CELLS

Graphene oxide and gold nanoparticles-based electrode has been used for oxygen reduction reaction in proton exchange membrane fuel cell. This electrode showed long-term stability and an efficient material for oxygen reduction. Traditional platinum-based electrodes were not able to satisfy all the requirements for oxygen reduction reaction. Researches relating to the effect of using the graphene oxide and gold nanoparticles-based cathodic electrode and as an innovative catalytic material are still in progress. As gold nanoparticles have high chemical reactivity and conductivity and large surface to volume ratio they are being explored for this application extensively [93].

8.8 CONCLUSION

Gold nanoshells as discussed above are being used widely in the medical field. We have seen above that its major usage is in cancer treatment. Cancer is a deadly disease and can be cured completely if diagnosed at proper time. So in the field of cancer treatment, these gold nanoshells are turning out to be a boon. Gold nanoshells are also being explored for various applications in the energy sectors, such as fuel cells, hydrogen storage, etc. The major drawback of these nanoshells is their cost. Researchers are still exploring more areas of applications for the gold nanoshells. In future, it might turn out to be the most commonly used nanomaterial in various fields and its cost would not restrict its usage.

KEYWORDS

- **nanoshells**
- **cores**
- **properties**
- **nanoshells**
- **plasmon hybridization**

REFERENCES

1. Kalele, S., Gosavi, S.W., Urban, J., Kulkarni, S.K. Nanoshell particles: Synthesis, properties and applications. *Current Science* 2006, 91(8).
2. Dahman, Y. *Nanoshells, Nanotechnology and Functional Materials for Engineers* 2017.
3. Loo, C., Lin, A., Hirsch, L., Lee, M.H., Barton, J., Halas, N., West, J., Drezek, R. Nanoshell-enabled photonics-based imaging and therapy of cancer. *Technology in Cancer Research & Treatment* 2004, 3(1), 33–40.
4. Brinson, B.E., Lassiter, J.B., Levin, C.S., Bardhan, R., Mirin, N., Halas, N.J. Nanoshells made easy: Improving Au layer growth on nanoparticle surfaces. *Langmuir* 2008, 24 (24), 14166–14171.
5. https://en.wikipedia.org/wiki/Nanoshell.
6. Wang, Y.C., Rheaume, E., Lesage, F., Kakkar, A. Synthetic methodologies to gold nanoshells: An overview. *Molecules* 2018, 23(11), 2851.
7. Zhou, H.S., Honma, I., Komiyama, H., Haus, J.W. Controlled synthesis and quantum-size effect in gold-coated nanoparticles. *Physical Review B* 1994, 50, 12052–12056.
8. Oldenburg, S.J., Averitt, R.D., Westcott, S.L., Halas, N.J. Nanoengineering of optical resonances. *Chemical Physics Letter* 1998, 288, 243–247.
9. Carpin, L.B., Bickford, L.R., Agollah, G., Yu, T.-K., Schiff, R., Li, Y., Drezek, R.A. Immunoconjugated gold nanoshell-mediated photothermal ablation of trastuzumab-resistant breast cancer cells. *Breast Cancer, Research and Treatment* 2011, 125, 27–34.
10. Qu, Y. Porter, R. Shan, F. Carter, J.D. Guo, T. Synthesis of tubular gold and silver nanoshells using silica nanowire core templates. *Langmuir* 2006, 22, 6367–6374.
11. Pol, V.G., Gedanken, A., Calderon-Moreno, J. Deposition of gold nanoparticles on silica spheres: A sonochemical approach. *Chemical Materials* 2003, 15, 1111–1118.
12. Suslick, K.S. *Ultrasound: Its Chemical, Physical, and Biological Effects*. VCH Publishers: Weinheim, Germany, 1988.
13. Zhang, Q., Ge, J., Goebl, J., Hu, Y., Sun, Y., Yin, Y. Tailored synthesis of superparamagnetic gold nanoshells with tunable optical properties. *Advanced Materials* 2010, 22, 1905–1909.
14. Zhang, Q., Zhang, T., Ge, J., Yin, Y. Permeable silica shell through surface-protected etching. *Nano Letter* 2008, 8, 2867–2871.
15. Storti, B., Elisei, F., Abbruzzetti, S., Viappiani, C., Latterini, L. One-pot synthesis of gold nanoshells with high photon-to-heat conversion efficiency. *Journal of Physical Chemistry, C* 2009, 113, 7516–7521.
16. Wang, X., Lin, K.S.K., Chan, J.C.C., Cheng, S. Preparation of ordered large pore SBA-15 silica functionalized with aminopropyl groups through one-pot synthesis. *Chemical Communications* 2004, 23, 2762–2763.
17. Zhang, J., Liu, J., Wang, S., Zhan, P., Wang, Z., Ming, N. Facile methods to coat polystyrene and silica colloids with metal. *Advanced Functional Materials* 2004, 14, 1089–1096.
18. Lee, J.-H., Mahmoud, M.A., Sitterle, V.B., Sitterle, J.J., Meredith, J.C. Highly scattering, surface-enhanced Raman scattering-active, metal nanoparticle-coated

polymers prepared via combined swelling-heteroaggregation. *Chemical Materials* 2009, 21, 5654–5663.
19. Dokoutchaev, A., James, J.T., Koene, S.C., Pathak, S., Prakash, G.K.S., Thompson, M.E. Colloidal metal deposition onto functionalized polystyrene microspheres. *Chemical Materials* 1999, 11, 2389–2399.
20. Shi, W., Sahoo, Y., Swihart, M.T., Prasad, P.N. Gold nanoshells on polystyrene cores for control of surface plasmon resonance. *Langmuir* 2005, 21, 1610–1617.
21. Liu, X., Gao, C., Gu, J., Jiang, Y., Yang, X., Li, S., Gao, W., An, T., Duan, H., Fu, J., et al. Hyaluronic acid stabilized iodine-containing nanoparticles with Au nanoshell coating for X-ray CT imaging and photothermal therapy of tumors. *ACS Applied Materials and Interfaces* 2016, 8, 27622–27631.
22. Bao, H., Bihr, T., Smith, A.-S., Klupp Taylor, R.N. Facile colloidal coating of polystyrene nanospheres with tunable gold dendritic patches. *Nanoscale* 2014, 6, 3954–3966.
23. Graf, C., van Blaaderen, A. Metallodielectric colloidal core–shell particles for photonic applications. *Langmuir* 2002, 18, 524–534.
24. Liang, H.-P., Wan, L.-J., Bai, C.-L., Jiang, L. Gold hollow nanospheres: Tunable surface plasmon resonance controlled by interior-cavity sizes. *Journal of Physical Chemistry, B* 2005, 109, 7795–7800.
25. Glavee, G.N., Klabunde, K.J., Sorensen, C.M., Hadjipanayis, G.C. Borohydride reduction of cobalt ions in water, chemistry leading to nanoscale metal, boride, or borate particles. *Langmuir* 1993, 9, 162–169.
26. Sugimoto, T. The theory of the nucleation of monodisperse particles in open systems and its application to AgBr systems. *Journal of Colloid and Interface Science* 1992, 150, 208–225.
27. Matijevic, E. Preparation and properties of uniform size colloids. *Chemical Materials* 1993, 5, 412–426.
28] Adams, S., Thai, D., Mascona, X., Schwartzberg, A.M., Zhang, J.Z. Key factors affecting the reproducibility of synthesis and growth mechanism of near-infrared absorbing hollow gold nanospheres. *Chemical Materials* 2014, 26, 6805–6810.
29. Lin, A.Y., Ariel, J.K.Y., Nixon, V., Drezek, R.A. Encapsulated Fe3O4/Ag complexed cores in hollow gold nanoshells for enhanced theranostic magnetic resonance imaging and photothermal therapy. *Small* 2014, 10, 3246–3251.
30. Sun, Y., Mayers, B., Xia, Y. Metal nanostructures with hollow interiors. *Advanced Materials* 2003, 15, 641–646.
31. Sun, Y., Mayers, B.T., Xia, Y. Template-engaged replacement reaction: A one-step approach to the large-scale synthesis of metal nanostructures with hollow interiors. *Nano Letters* 2002, 2, 481–485.
32. Hu, T., Lin, Y., Yan, J., Di, J. Synthesis of hollow gold nanoparticles on the surface of indium tin oxide glass and their application for plasmonic biosensor. *Spectrochimica, Acta Part A* 2013, 110, 72–77.
33. Huang, D., Hu, T., Chen, N., Zhang, W., Di, J. Development of silver/gold nanocages onto indium tin oxide glass as a reagentless plasmonic mercury sensor. *Analytica Chimica Acta* 2014, 825, 51–56.

34. Yan, Z., Hu, T., Guo, W., Deng, A., Di, J. A label-free immunosensor for determination of salbutamol based on localized surface plasmon resonance biosensing. *Bioprocess and Biosystems Engineering* 2014, 37, 651–657.
35. Radloff, C., Vaia, R.A., Brunton, J., Bouwer, G.T., Ward, V.K. Metal nanoshell assembly on a virus bioscaffold. *Nano Letter* 2005, 5, 1187–1191.
36. Dujardin, E., Peet, C., Stubbs, G., Culver, J.N., Mann, S. Organization of metallic nanoparticles using tobacco mosaic virus templates. *Nano Letter* 2003, 3, 413–417.
37. Behrens, S., Rahn, K., Habicht, W., Bohm, K.-J., Rosner, H., Dinjus, E., Unger, E. Nanoscale particle arrays induced by highly ordered protein assemblies. *Advanced Materials* 2002, 14, 1621–1625.
38. Prodan, E., Radloff, C., Halas, N.J., Nordlander, P. A hybridization model for the plasmon response of complex nanostructures. *Science* 2003, 302, 419–422.
39. Bardhan, R., Mukherjee, S., Mirin, N.A., Levit, S.D., Nordlander, P., Halas, N.J. Nanosphere-in-a-nanoshell: A simple nanomatryushka. *Journal of Physical Chemistry, C* 2010, 114, 7378–7383.
40. Xiaohu, X., Yang, L., Vadim, B., Guillermo, A.A. Engineering sub-100 nm multi-layer nanoshells. *Nanotechnology* 2006, 17, 5435.
41. Yongping, G., Yongsheng, L., Yao, W., Yi, C., Jinlou, G., Wenru, Z., Jian, D., Jianlin, S. Controlled synthesis of multilayered gold nanoshells for enhanced photothermal therapy and SERS detection. *Small* 2015, 11, 77–83.
42. Chen, H.M., Liu, R.-S., Asakura, K., Lee, J.-F., Jang, L.-Y., Hu, S.-F. Fabrication of nanorattles with passive Shell. *Journal of Physical Chemistry, B* 2006, 110, 19162–19167.
43. Mazzucco, S., Stéphan, O., Colliex, C., Pastoriza-Santos, I., Liz-Marzan, L.M., de Abajo, J.G., Kociak, M. Spatially resolved measurements of plasmonic eigenstates in complex-shaped, asymmetric nanoparticles: Gold nanostars. *European Physical Journal Applied Physics* 2011, 54, 33512.
44. Gersten, J., Nitzan, A. Electromagnetic theory of enhanced Raman scattering by molecules adsorbed on rough surfaces. *Journal of Physical Chemistry* 1980, 73, 3023–3037.
45. Rodríguez-Lorenzo, L., de la Rica, R., Álvarez-Puebla, R.A., Liz-Marzán, L.M., Stevens, M.M. Plasmonic nanosensors with inverse sensitivity by means of enzyme-guided crystal growth. *Nature Materials* 2012, 11, 604.
46. Ray, P.C. Size and shape dependent second order nonlinear optical properties of nanomaterials and their application in biological and chemical sensing. *Chemical Review* 2010, 110, 5332–5365.
47. Sanchez-Gaytan, B.L., Park, S.-J. Spiky gold nanoshells. *Langmuir* 2010, 26, 19170–19174.
48. Sanchez-Gaytan, B.L., Swanglap, P., Lamkin, T.J., Hickey, R.J., Fakhraai, Z., Link, S., Park, S.-J. Spiky gold nanoshells: Synthesis and enhanced scattering properties. *Journal of Physical Chemistry C* 2012, 116, 10318–10324.
49. Pedireddy, S., Li, A., Bosman, M., Phang, I.Y., Li, S., Ling, X.Y. Synthesis of spiky Ag–Au octahedral nanoparticles and their tunable optical properties. *Journal of Physical Chemistry C* 2013, 117, 16640–16649.
50. Topete, A., Alatorre-Meda, M., Iglesias, P., Villar-Alvarez, E.M., Barbosa, S., Costoya, J.A., Taboada, P., Mosquera, V. Fluorescent drug-loaded, polymeric-based,

branched gold nanoshells for localized multimodal therapy and imaging of tumoral cells. *American Chemical Society Nano* 2014, 8, 2725–2738.

51. Topete, A., Alatorre-Meda, M., Villar-Álvarez, E.M., Cambón, A., Barbosa, S., Taboada, P., Mosquera, V. Simple control of surface topography of gold nanoshells by a surfactant-less seeded-growth method. *American Chemical Society Applied Materials & Interfaces* 2014, 6, 11142–11157.
52. Bai, X., Wang, Y., Song, Z., Feng, Y., Chen, Y., Zhang, D., Feng, L. The basic properties of gold nanoparticles and their applications in tumor diagnosis and treatment. *International Journal of Molecular Science* 2020, 21(7), 2480.
53. Jauffred, L., Samadi, A., Klingberg, H., Bendix, P.M., Oddershede, L.B. Plasmonic heating of nanostructures. *Chemical Reviews* 2019, 119, 8087–8130.
54. Schuller, J.A., Barnard, E.S., Cai, W., Jun, Y.C., White, J.S., Brongersma, M.L. Plasmonics for extreme light concentration and manipulation. *Nature Materials* 2010, 9, 193–204.
55. Seltzer, S.M., Tables of x-ray mass attenuation coefficients and mass energy–absorption coefficients NISTIR 5632. *Pramana* 1995, 72, 375–387.
56. Her, S., Jaffray, D., Allen, C. Gold nanoparticles for applications in cancer radiotherapy: Mechanisms and recent advancements. *Advanced Drug Delivery Reviews* 2017, 109, 84–101.
57. Curran, W.J., Littman, P., Raney, R.B. Interstitial radiation therapy in the treatment of childhood soft-tissue sarcomas. *International Journal of Radiation Oncology* 1988, 14, 169–174.
58. Guo, Y., Aweda, T., Black, K.C., Liu, Y. Chemistry and theranostic applications of radiolabeled nanoparticles for cardiovascular, oncological, and pulmonary research. *Current Topics in Medicinal Chemistry* 2013, 13, 470–478.
59. Jans, H., Huo, Q. Gold nanoparticle-enabled biological and chemical detection and analysis. *Chemical Society Reviews* 2012, 41, 2849–2866.
60. Hainfeld, J. A small gold-conjugated antibody label: Improved resolution for electron microscopy. *Science* 1987, 236, 450–453.
61. Comotti, M., Della Pina, C., Matarrese, R., Rossi, M. The catalytic activity of "Naked" gold particles. *Angewandte Chemie International Edition* 2004, 43, 5812–5815.
62. Jv, Y. Li, B., Cao, R. Positively-charged gold nanoparticles as peroxidiase mimic and their application in hydrogen peroxide and glucose detection. *Chemical Communications* 2010, 46, 8017.
63. Bhattacharya, R., Patra, C.R., Verma, R., Kumar, S., Greipp, P.R., Mukherjee, P. Gold nanoparticles inhibit the proliferation of multiple myeloma cells. *Advanced Materials* 2007, 19, 711–716.
64. Cai, W., Gao, T., Hong, H., Sun, J. Applications of gold nanoparticles in cancer nanotechnology. *Nanotechnology Science Applications* 2008, 1, 17–32
65. Van Horssen, R., Ten Hagen, T.L., Eggermont, A.M. TNF-alpha in cancer treatment: Molecular insights, antitumor effects, and clinical utility. *Oncologist* 2006, 11, 397–408.
66. Wang, S., Lu, G. Applications of gold nanoparticles in cancer imaging and treatment. In *Noble and Precious Metals: Properties, Nanoscale Effects and Applications*. IntechOpen, 2017.

67. Braun, G.B., Pallaoro, A., Wu, G., Missirlis, D., Zasadzinski, J.A., Tirrell, M., et al. Laser-activated gene silencing via gold nanoshell-siRNA conjugates. *American Chemical Society Nano* 2009, 3(7), 2007–2015.
68. Jensen, S.A., Day, E.S., Ko, C.H., Hurley, L.A., Luciano, J.P., Kouri, F.M., et al. Spherical nucleic acid nanoparticle conjugates as an RNAi-based therapy for glioblastoma. *Science Translational Medicine* 2013, 5(209), 209ra152.
69. Wang, S., Tian, Y., Tian, W., Sun, J., Zhao, S., Liu, Y., et al. Selectively sensitizing malignant cells to photothermal therapy using a CD44-targeting heat shock protein 72 depletion nano-system. *American Chemical Society Nano* 2016, 10(9), 8578–8590.
70. Lin, A.Y., Lunsford, J., Bear, A.S., Young, J.K., Eckels, P., Luo, L., et al. High-density sub-100-nm peptide-gold nanoparticle complexes improve vaccine presentation by dendritic cells in vitro. *Nanoscale Research Letters* 2013, 8(1), 72.
71. Lee, I.-H., Kwon, H.-K., An, S., Kim, D., Kim, S., Yu, M.K. et al. Imageable antigen-presenting gold nanoparticle vaccines for effective cancer immunotherapy in vivo. *Angewandte Chemie, International Edition* 2012, 51(35), 8800–8805.
72. Sokolov, K., Follen, M., Aaron, J., Pavlova, I., Malpica, A., Lotan, R., et al. Real-time vital optical imaging of precancer using anti-epidermal growth factor receptor antibodies conjugated to gold nanoparticle. *Cancer Research* 2003, 63(9), 1999–2004.
73. El-Sayed, I.H., Huang, X., El-Sayed, M.A. Surface plasmon resonance scattering and absorption of anti-EGFR antibody conjugated gold nanoparticles in cancer diagnostics: Applicacations in oral cancer. *Nano Letters* 2005, 5(5), 829–834.
74. Gobin, A.M., Lee, M.H., Halas, N.J., James, W.D., Drezek, R.A., West, J.L. Near-infrared resonant nanoshells for combined optical imaging and photothermal cancer therapy. *NanoLetters* 2007, 7(7), 1929–1934.
75. Nie, L., Wang, S., Wang, X., Rong, P., Ma, Y., Liu, G., et al. In vivo volumetric photoacoustic molecular angiography and therapeutic monitoring with targeted plasmonic nanostars. *Small* 2014, 10(8), 1585–1593.
76. Tian, Y., Zhang, Y., Teng, Z., Tian, W., Luo, S., Kong, X., et al. pH-dependent transmembrane activity of peptide-functionalized gold nanostars for computed tomography/photoacoustic imaging and photothermal therapy. *American Chemical Society, Applied Materials & Interfaces* 2017, 9(3), 2114–2122.
77. Aydogan, B., Li, J., Rajh, T., Chaudhary, A., Chmura, S.J., Pelizzari, C., et al. AuNP-DG:Deoxyglucose-labeled gold nanoparticles as X-ray computed tomography contrast agents for cancer imaging. *Molecular Imaging and Biology* 2010, 12(5), 463–467.
78. Wang, H., Zheng, L., Peng, C., Guo, R., Shen, M., Shi, X., et al. Computed tomography imaging of cancer cells using acetylated dendrimer-entrapped gold nanoparticles. *Biomaterials* 2011, 32(11), 2979–2988.
79. Tian, Y., Luo, S., Yan, H., Teng, Z., Pan, Y., Zeng, L., et al. Gold nanostars functionalized with amine-terminated PEG for X-ray/CT imaging and photothermal therapy. *Journal of Materials Chemistry B* 2015, 3, 4330–4337.
80. Hainfeld, J.F., O'Connor, M.J., Dilmanian, F.A., Slatkin, D.N., Adams, D.J., Smilowitz, H.M. Micro-CT enables microlocalisation and quantification of Her2-targeted gold nanoparticles within tumour regions. *The British Journal of Radiology* 2011, 84(1002), 526–533.

81. Jain, S., Hirst, D.G., Sullivan, M.O. Gold nanoparticles as novel agents for cancer therapy. *The British Journal of Radiology* 85, 2012, 101–113.
82. Wust, P., Hildebrandt, B., Sreenivasa, G., Rau, B., Gellermann, J., Riess, H., et al. Hyperthermia in combined treatment of cancer. *Lancet Oncology* 2002, 3, 487–497.
83. Lal, S., Clare, S.E., Halas, N.J. Nanoshell-enabled photothermal cancer therapy, impending clinical impact. *Accounts of Chemical Research* 2008, 41, 1842–1851.
84. Hirsch, L., Stafford, R., Bankson, J., Sershen, S., Rivera, B., Price, R. et al. Nanoshell-mediated near-infrared thermal therapy of tumors under magnetic resonance guidance. *Proceedings of the National Academy of Sciences of the United States of America* 2003, 100, 13549–13554.
85. Stern, J.M., Stanfield, J., Lotan, Y., Park, S., Hsieh, J.T., Cadeddu, J.A. Efficacy of laser-activated gold nanoshells in ablating prostate cancer cells in vitro. *Journal of Endurology* 2007, 21, 939–943.
86. Haume, K., Rosa, S., Grellet, S., Śmiałek, M.A., Butterworth, K.T., Solov'yov, A.V., et al. Gold nanoparticles for cancer radiotherapy: A review. *Cancer Nanotechnology* 2016, 7(1), 8.
87. Hainfeld, J.F., Slatkin, D.N., Smilowitz, H.M. The use of gold nanoparticles to enhance radiotherapy in mice. *Physics in Medicine & Biology* 2004, 49(18), N309–N315.
88. Geng, F., Song, K., Xing, J.Z., Yuan, C., Yan, S., Yang, Q., et al. Thio-glucose bound gold nanoparticles enhance radio-cytotoxic targeting of ovarian cancer. *Nanotechnology* 2011, 22(28), 285101.
89. Chattopadhyay, N., Cai, Z., Kwon, Y.L., Lechtman, E., Pignol, J.-P., Reilly, R.M. Molecularly targeted gold nanoparticles enhance the radiation response of breast cancer cells and tumor xenografts to X-radiation. *Breast Cancer Research and Treatment* 2013, 137(1), 81–91.
90. Zhao, J., Wallace, M., Melancon, M.P. Cancer theranostics with gold nanoshells. *Nanomedicine*, 2014, 9(13), 2041–2057.
91. Kohout, C., Santi, C., Polito, L. Anisotropic gold nanoparticles in biomedical applications. *International Journal of Molecular Science* 2018, 19, 11, 3385.
92. Anjana, P.M., Bindhu, M.R., Rakhi, R.B. Green synthesized gold nanoparticle dispersed porous carbon composites for electrochemical energy storage. *Materials Science for Energy Technologies* 2019, 2(3), 389–395.
93. Marinoiu, A., Andrulevicius, M., Tamuleviciene, A., Tamulevicius, T., Raceanu, M., Varlam, M. Synthesis of well dispersed gold nanoparticles on reduced graphene oxide and application in PEM fuel cells. *Applied Surface Science* 2020, 504, 14451

CHAPTER 9

Design and Modeling of an Invelox Wind Turbine: A Nonconventional Energy Source

HIRANMOY SAMANTA,[1*] JOYDIP PAUL,[1] TANUJIT METE,[2] and KAMAL GOLUI[1]

[1]*Department of Mechanical Engineering, Gargi Memorial Institute of Technology, Balarampur, Baruipur, Kolkata, West Bengal, India*

[2]*Department of Mechanical Engineering, Kingston Polytechnic College, Kolkata, West Bengal, India*

*Corresponding author

ABSTRACT

This chapter explores the key principles and advantages of the Invelox wind turbine technology. An Invelox wind turbine is an innovative wind energy technology that diverges from traditional wind turbine designs by capturing and harnessing wind energy into a funnel-like structure, which significantly increases wind speed toward turbine leads to enhancing the overall efficiency of power generation. Invelox wind turbine adoption, emphasizing the technology's role in contributing to sustainable and clean energy solutions. By maintaining the aspect ratio, an experimental setup was created, and tests were run to determine the flow parameters. The computational fluid dynamics model has been developed based on *finite volume method* and the presentation of numerical data. To improve calculations and examine the relationship between the inner geometry of the system and the direction of the wind, changes were made. The experimental set up evident the validation of the physical model. The numerical

Advances in Energy Materials: Materials Fabrication, Properties and Applications.
Lishin T, Maciej Jaroszewski, and Sabu Thomas (Eds.)

DOI: 10.1201/9781779643674-9

model represents the different configuration with a specified angle on the vertical plane at 0° and 180°. The analysis was done on ANSYS with the 6-lakh grid. Velocity and pressure contours are presented in the chapter.

9.1 INTRODUCTION

The innovation in the Invelox design, which differentiates it from other conventional turbine, is to collect air from all directions and direct it through a venture section toward the turbine blades. The conventional turbines are mainly vertical or horizontal whereas the Invelox is omnidirectional and considering the same wind conditions and parameters it delivers significantly more higher output at the same time at a reduced cost. The INVELOX design perspective catches the impact of the high velocity wind energy generation by striking the turbine blades by directing the wind omni-directionally and directed toward the funnel-shaped design led to the ground level. By maintaining the aspect ratio, an experimental setup was created, and tests were run to determine the flow parameters. The development of a computational fluid dynamics (CFD) model is mainly based on the finite volume method. The presentation of numerical data and figures reflected the characteristic and flow pattern. The presentation of numerical data and figures reflected the characteristic and flow pattern. To improve the calculations and examine the relationship between the inner geometry of the system and the direction of the wind, changes were made. The study demonstrates that it is possible

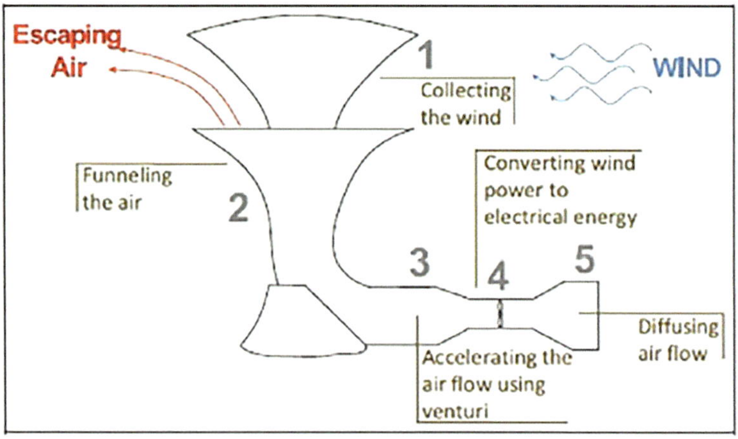

FIGURE 9.1 Schematic of INVELOX delivery system.

to direct wind flow by capturing, accelerating, and concentrating at a particular optimization level in order to provide the most power. The FVM approach is applied. The modeling is represented by ANSYS platform ids.

Ghorani et al. [1] have studied and analyzed the geometry of the Invelox wind using different types of computational and optimization-based methods. The comparison of Invelox with HAWT has been presented in this chapter. The performance of INVELOX has been studied and enhanced by Farokhzade and Maghrebi [2] for various wind directions and also by adjusting the number of blades, while wind escaping can be decreased by lengthening the upper funnel. According to a study by Ahmed et al. [3], increased wind velocity increases output power. Aerodynamic adjustable roof structures help to increase efficiency while no wind is fleeing, the system was researched by Golozar et al. [7] in 2020. Roof orientations had been managed via fuzzy controls. The efficiency with which wind turbines may produce power in locations with low wind velocities was examined by Venkatramakrishnan et al. [5] in 2020. Li [4] and Li and Guo [6] used CFD to investigate the features of the flow field while taking the atmospheric boundary layer into account. Gohar et al. [8] investigated a range of INVELOX models to determine the relationship between power output and pressure and velocity contours. In order to numerically reduce the volume of exiting air in an INVELOX turbine, Anbarsooz et al. [9] have given three designs. Gavade et al. [10] performed numerous simulations and investigated the relationship between INVELOX shape, wind direction, and fluid flow. The venturi part of the INVELOX was examined and researched by Solanki et al. [11] in 2017. The power output is discovered to be increased by a multistage turbine in the venturi portion. By catching, accelerating, and focusing the wind, Allaci and Andreopoulos [12] investigated both internal and external fluid flow.

9.2 EXPERIMENTAL SETUP

The Invelox turbine with the right model to prototype ratio makes up the experimental setup (Figure 9.2). After a successful grid independence test, the numerical findings were verified using the experimental data. Additionally, the experimental results were verified by Abe and Ohya [13]. The numerical data and the experimental data exhibit high agreement. At various points in the setup, probes are employed to measure the wind pressure and speed. A data recorder and computer are attached to the thermocouples.

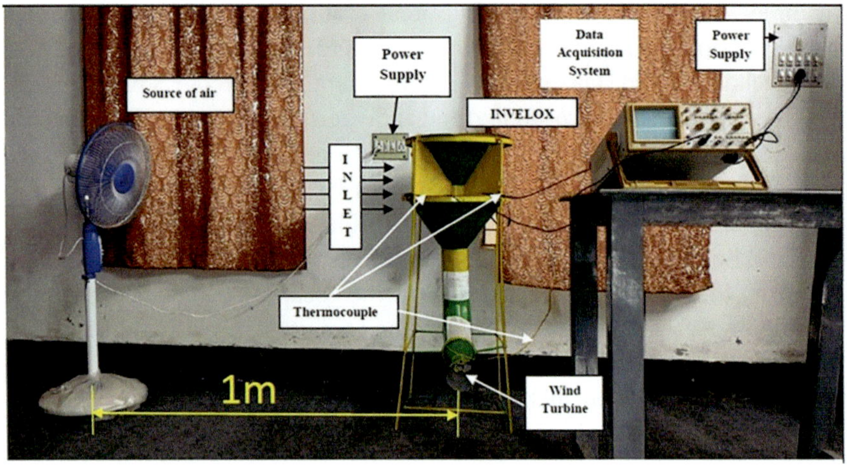

FIGURE 9.2 Experimental setup of Invelox turbine.

9.3 RESULT AND DISCUSSION

For the current investigation, numerical results have been derived and contrasted with the experimental findings of Abe and Ohya [13] (Figures 9.3 and 9.4). Turbulence model-derived velocity and pressure contours have been shown.

After successful grid independency test of 4-lakh, 6-lakh, and 8-lakh grid volume, the 6-lakh structured grid has been chosen.

The average venturi velocity is found to be affected by the k–ε turbulence model, and when it is compared to the numerical data of Allaei and Andreopoulos [12], there is a relative error of 0.9%, and the velocity is found to be 10.5 m/s.

The pressure contours of the Invelox in the vertical plane at 0° and 180° are shown in Figure 9.5. At zero, Section B has a higher average total pressure than Section A. Section B has a lower overall pressure at 180° than does Section A.

9.4 CONCLUSIONS

Using numerical simulation, the properties and design of Invelox are examined. The results are given as velocity and pressure contours. The numerical data and the numerical data of Abe and Ohya [13] have been compared.

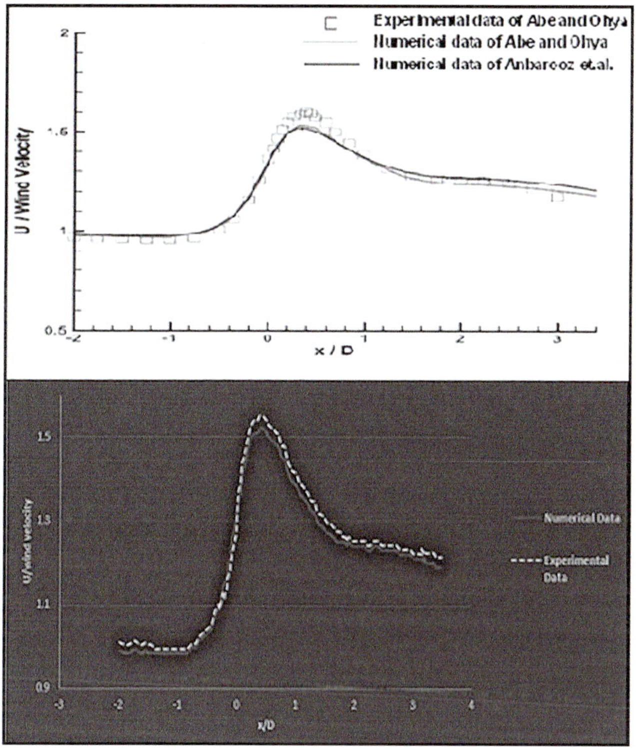

FIGURE 9.3 Comparison of numerical and experimental data with Abe and Ohya and Anbarsooz et al.

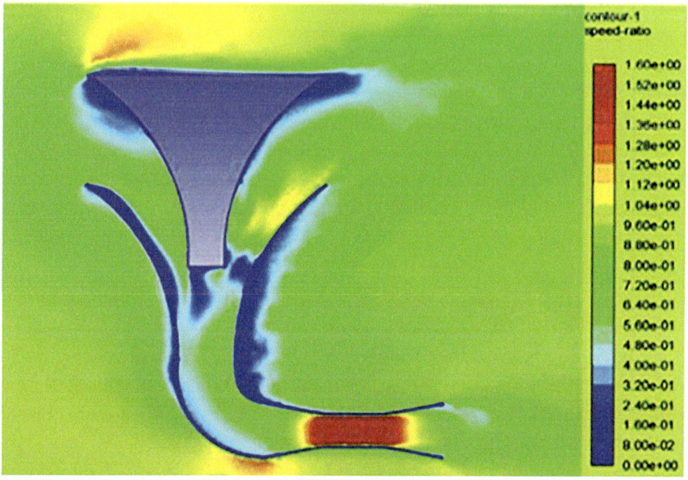

FIGURE 9.4 Standard k–ε velocity contour on median cross-section.

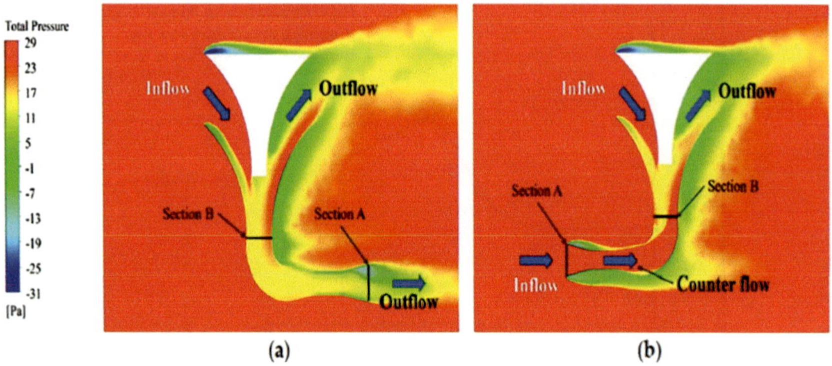

FIGURE 9.5 Pressure contours at (a) $\theta = 0°$ and (b) $\theta = 180°$.

KEYWORDS

- **INVELOX**
- **CFD**
- **finite volume method**
- **non convention wind turbine**
- **ANSYS**

REFERENCES

1. Ghorani, M.M., Karimi, B., Mirghavami, S.M., Saboohi, S. A numerical study on the feasibility of electricity production using an optimized wind delivery system (Invelox) integrated with a Horizontal axis wind turbine (HAWT). *Energy* 2023¸268, 126643.https://doi.org/10.1016/j.energy.2023.126643.
2. Farokhzade, A., Maghrebi, M.J. Inlet parameters effects of Invelox on the aerodynamic performance using numerical simulation. *Journal of Applied Fluid Mechanics* 14(5), 1511–1520.
3. Ahmed, A., A, A. Leevans. L, Astle. M, Sanju Edwards. M. "Design and fabrication of Invelox wind turbine. *International Research Journal of Engineering and Technology* 2021, 8(4) 1–6. | e-ISSN: 2395-0056| p-ISSN: 2395-0072.
4. Li, D. Study of Invelox wind turbine considering atmospheric boundary layer: Based on numerical simulation. *Journal of Physics: Conference Series* 2020, 1600, 012063.
5. Venkatramakrishnan, R., Pandey, J.K., Mondal, A.K., Karn, A. Low speed wind turbines for power generation: A review. *Journal of Advanced Research in Fluid Mechanics and Thermal Sciences* 2020, 67(1), 146–169.

6. Li, D., Guo, T. Numerical Study on the power efficiency and flow characteristics of a new type of wind energy collection device. *Applied Science* 2020, 10, 7438.
7. Golozar, A., Shirazi, F.A., Siahpour, S., Khakiani, F.N., Osguei, K.G. A novel aerodynamic controllable roof for improving performance of INVELOX wind delivery system. *Wind Engineering* 2021, 45(3), 477–490.
8. Gohar, G.A., Manzoor, T., Ahmad, A., Hameed, Z., Saleem, F., Ahmad, I., Sattar, A., Arshad, A. Design and comparative analysis of an INVELOX wind power generation system for multiple wind turbines through computational fluid dynamics. *Advances in Mechanical Engineering* 2019, 11(4), 1–10.
9. Anbarsooz, M., Amiri, M., Rashidi, I. A novel curtain design to enhance the aerodynamic performance of Invelox: A steady-RANS numerical simulation. *Energy* 2019, 168, 207–221.
10. Gavade, A.A., Mulla, A.S., Ransing, A.M., Sane, N.M. Design and manufacturing of INVELOX to generate wind power using nonconventional energy sources. *International Journal of Advance Research in Science and Engineering* 2018, 7(3), 2319–8354.
11. Solanki, A.L., Kayasth, B.D., Bhatt, H. Design modification & analysis for venturi section of INVELOX system to maximize power using multiple wind turbine. *International Journal for Innovative Research in Science & Technology* 2017, 3(11), 2349–6010.
12. Allaei, D., Andreopoulos, Y. INVELOX: Description of a new concept in wind power and its performance evaluation. *Energy* 2014, 69, 336–344.
13. Abe, K.I., Ohya, Y. An investigation of flow fields around flanged diffusers using CFD. *Journal of Wind Engineering and Industrial Aerodynamics* 2004, 92(3e4), 315e30.

Index

E

F

G

N

O

P

T

V

W

Z

For Product Safety Concerns and Information please contact our EU representative GPSR@taylorandfrancis.com Taylor & Francis Verlag GmbH, Kaufingerstraße 24, 80331 München, Germany